THANK YOU GOD!

Preschool Math Workbook

THANK YOU GOD! Preschool Math Workbook Copyright
© 2021 by Dr. Diana Carle
Nature's Living Classroom Publishing
For Information Contact:
Dr. Diana Carle at www.DoctorDianaCarle.com
Written and illustrated by Dr. Diana Carle

This Book Belongs to:

____

This workbook is designed to increase in difficulty by section so children can grow their confidence, and it requires guidance from the teacher to follow the exercises.

Section 1: Number Tracing

Section 2: Pattern Recognition

Section 3: Addition & Subtraction

Section 1: Number Tracing

We are going to learn to write our numbers and count some of God's creations in nature! Have fun and color God's creations as you count them!

Thank You God for the

1 duck that I can count!

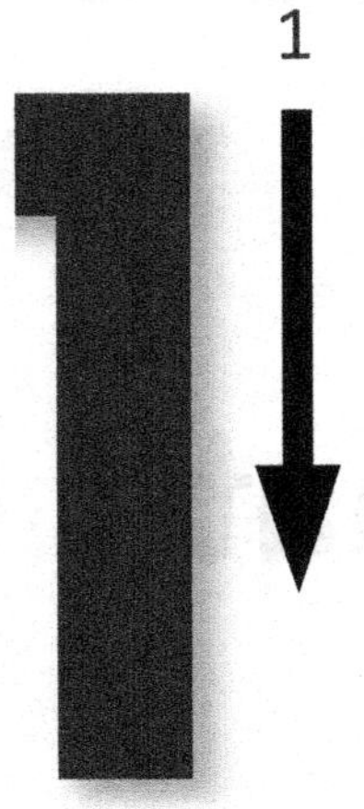
1

Thank You God for the

2 fawns that I can count!

2
1
2
3

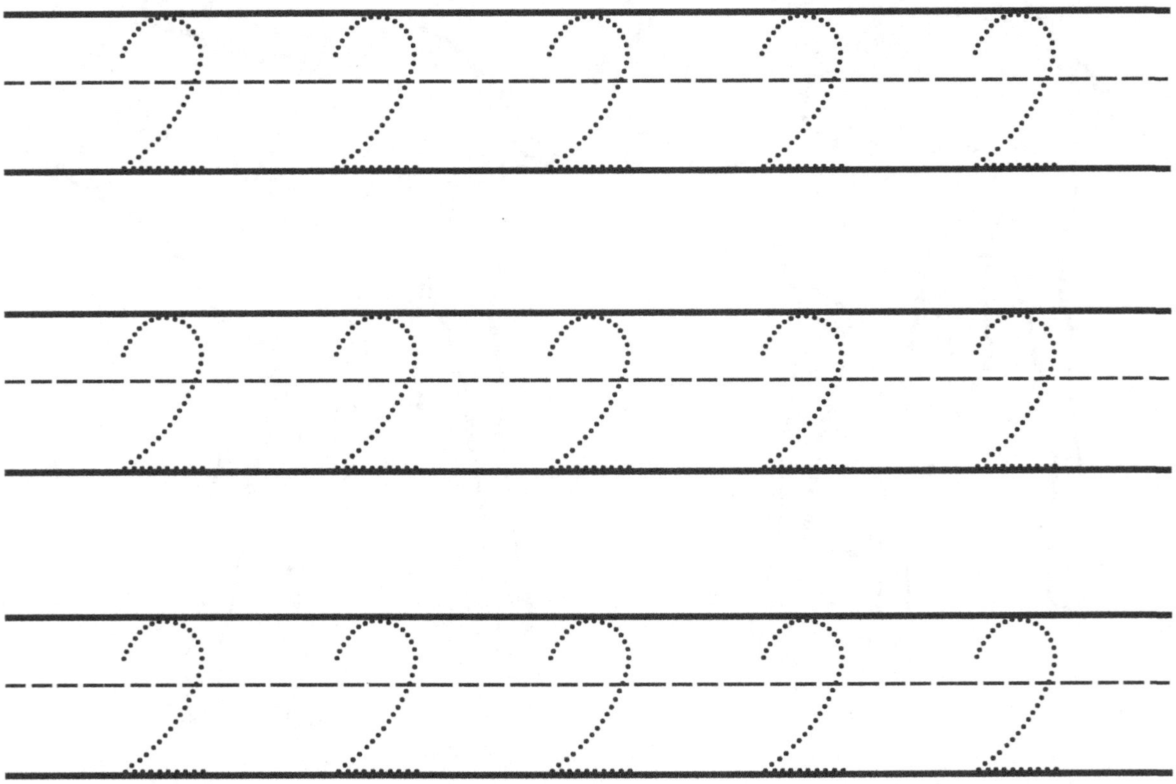

Thank You God for the

3 bunnies that I can count!

3 3 3 3 3

3 3 3 3 3

3 3 3 3 3

Thank You God for the

4 mice that I can count!

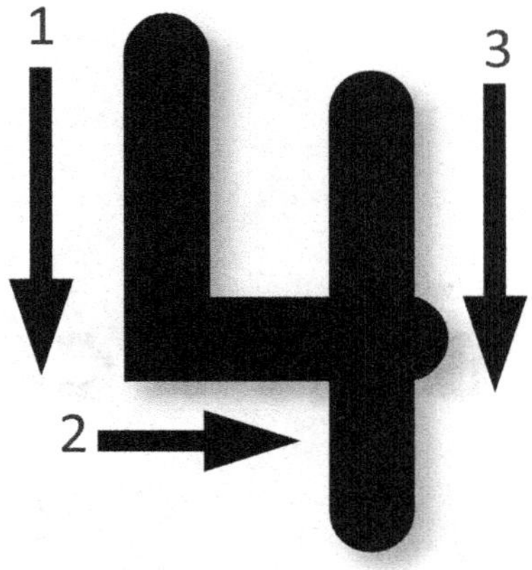
1
2
3

Thank You God for the

5 eagles that I can count!

1
2
3
5

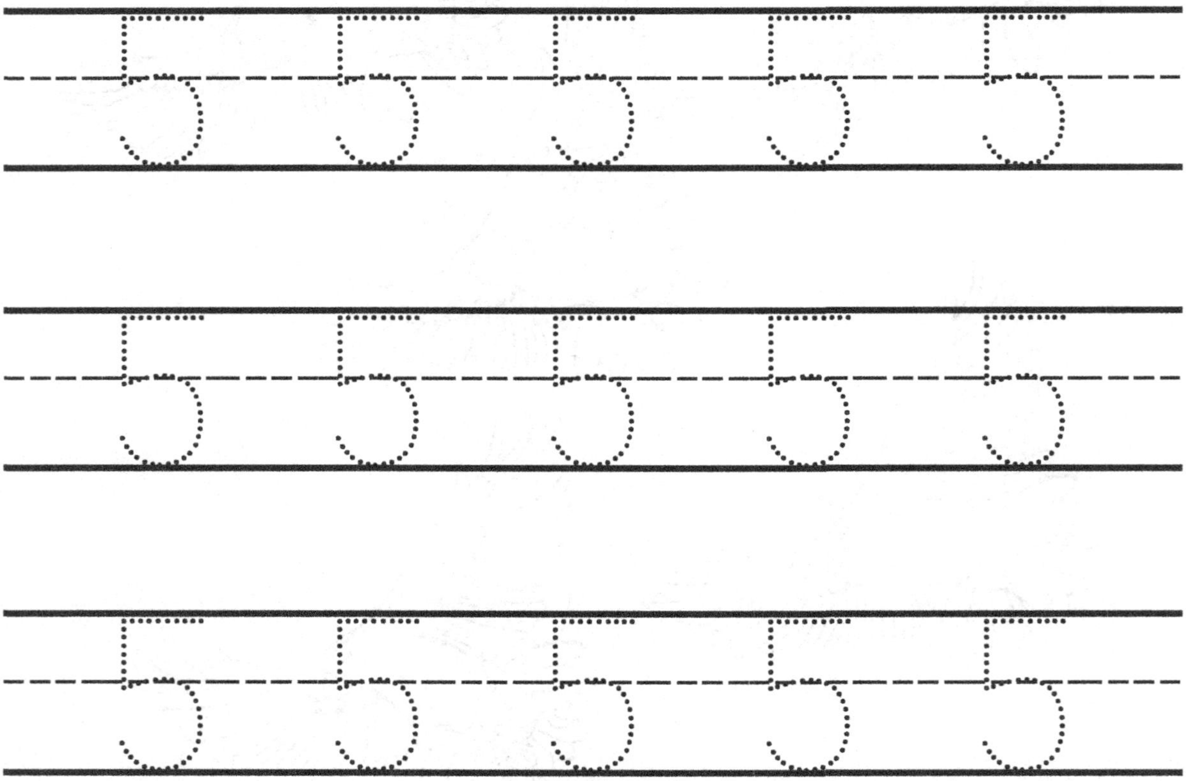

Thank You God for the

6 frogs that I can count!

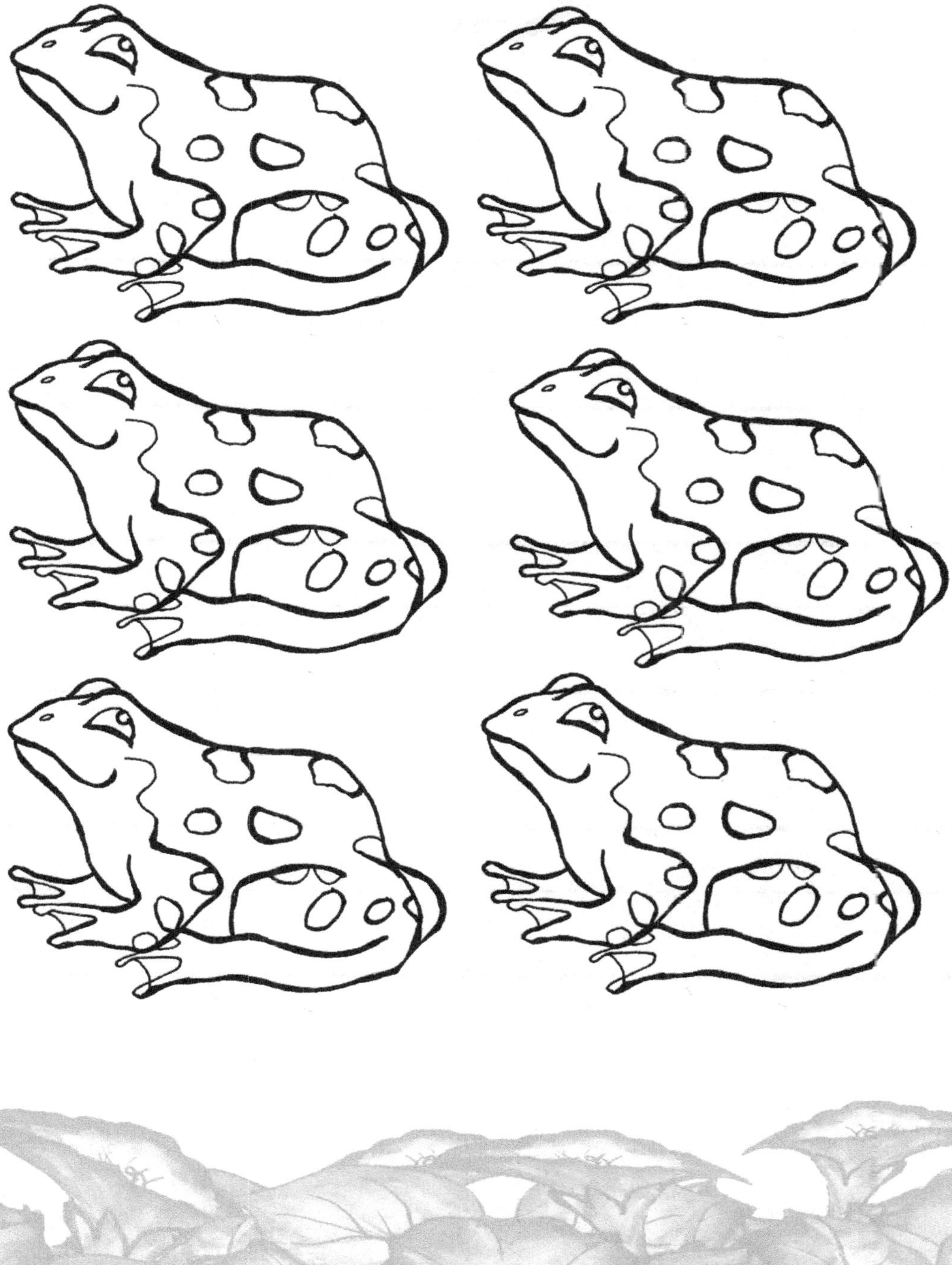

6 6 6 6 6

6 6 6 6 6

6 6 6 6 6

Thank You God for the

7 squirrels that I can count!

1
2

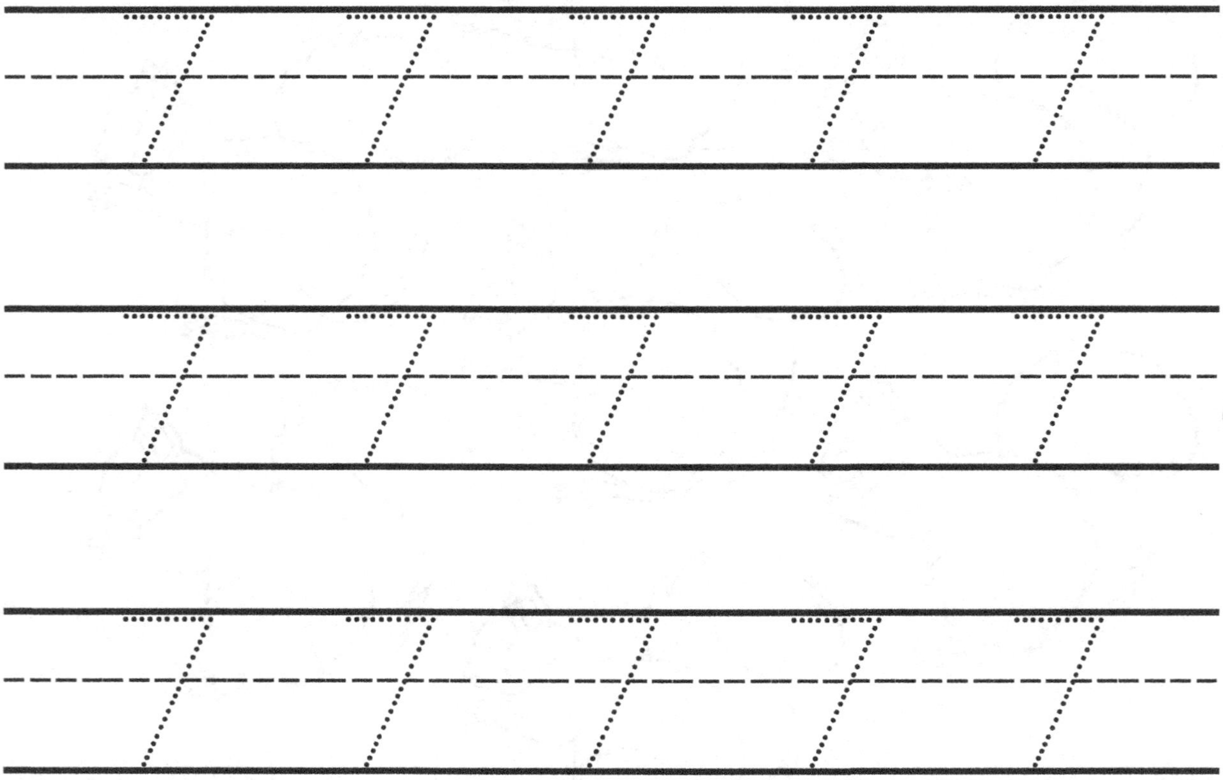

Thank You God for the

8 crickets that I can count!

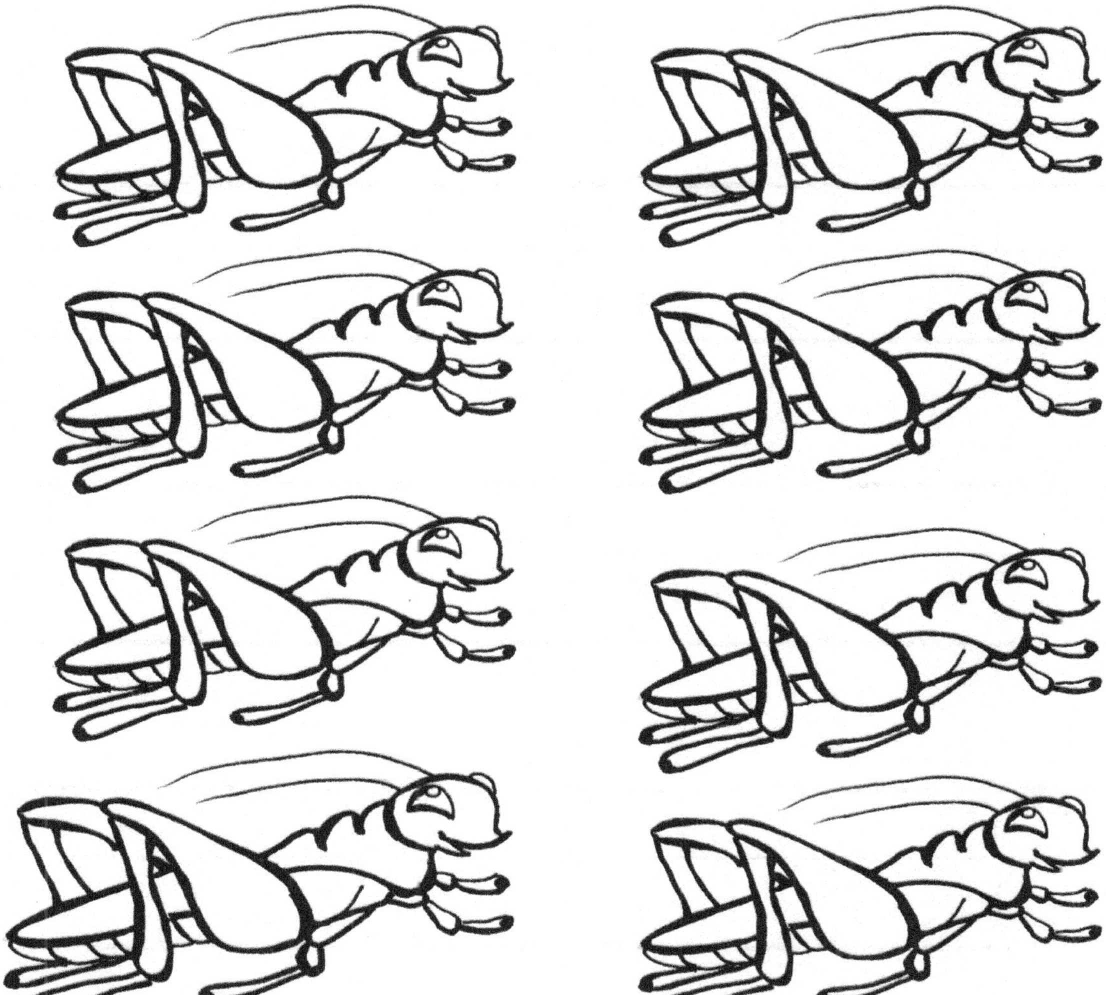

8
1
4
2
3

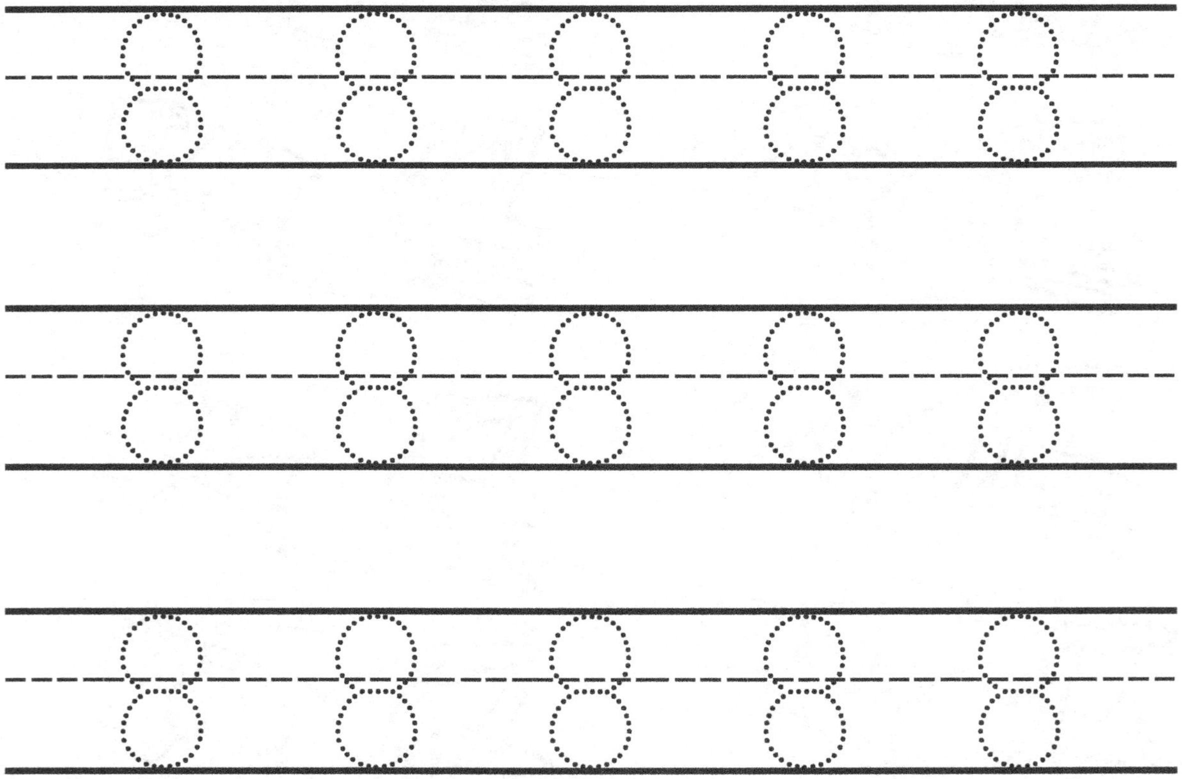

Thank You God for the

9 chipmunks that I can

count!

q q q q q

q q q q q

q q q q q

Thank You God for the

10 butterflies that I can

count!

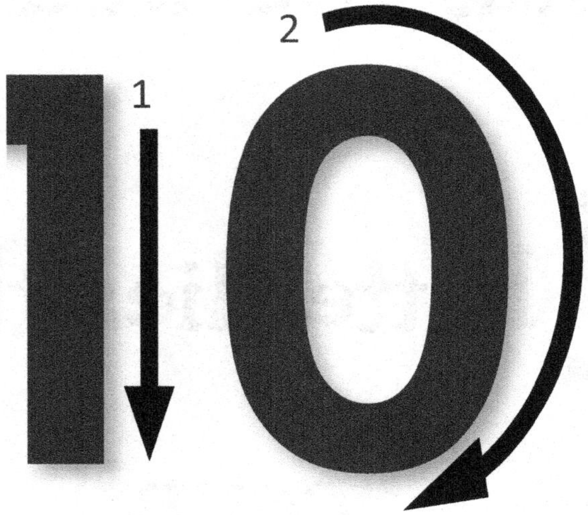

10 10 10 10

10 10 10 10

10 10 10 10

Thank You God for the

11 roses that I can count!

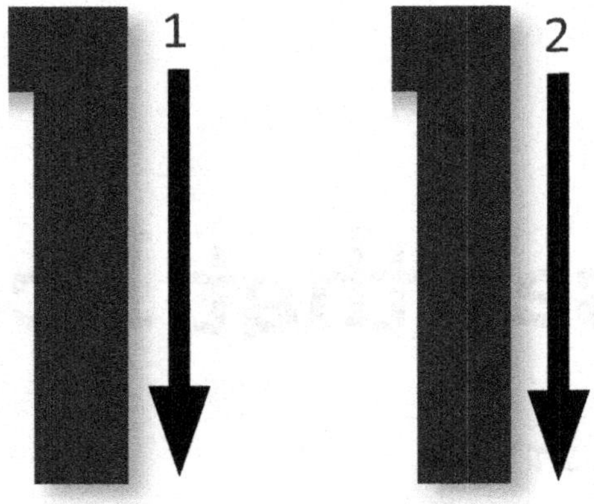
11
1
2

Thank You God for the

12 turtles that I can

count!

2 2 2 2

2 2 2 2

2 2 2 2

Thank You God for the

13 ladybug beetles that I

can count!

13
1
2
3

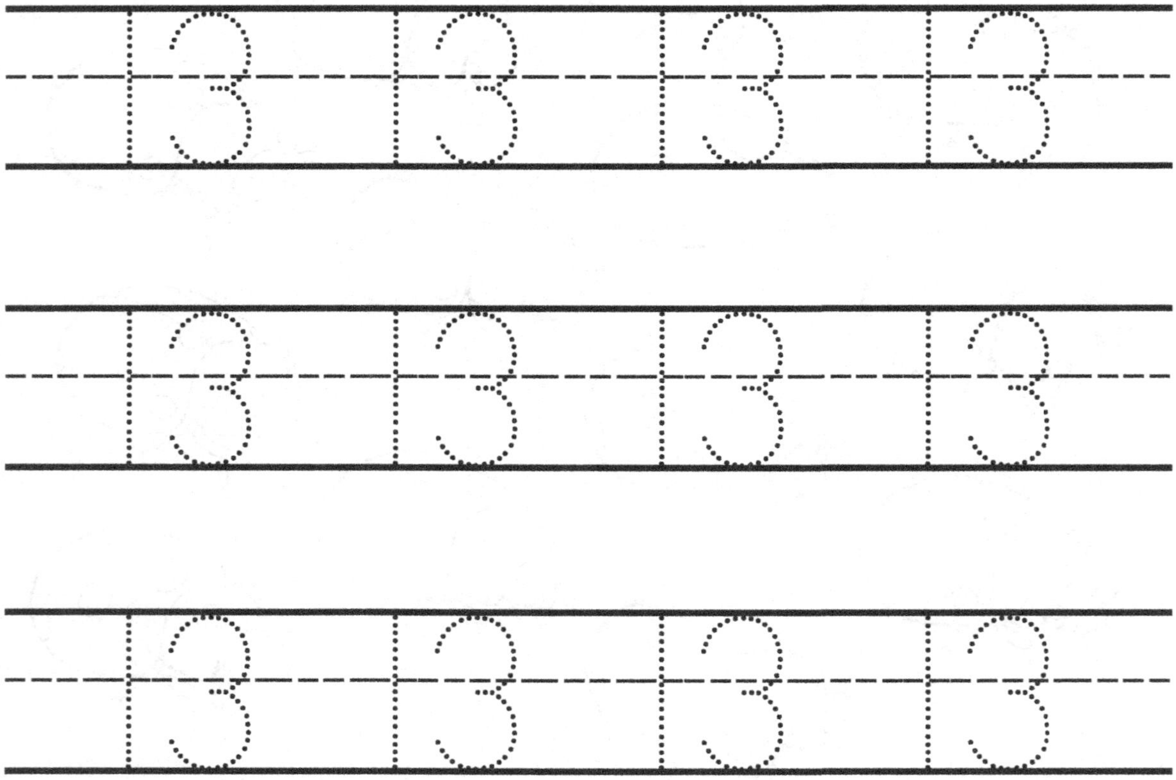
13 13 13 13
13 13 13 13
13 13 13 13

Thank You God for the

14 trout that I can count!

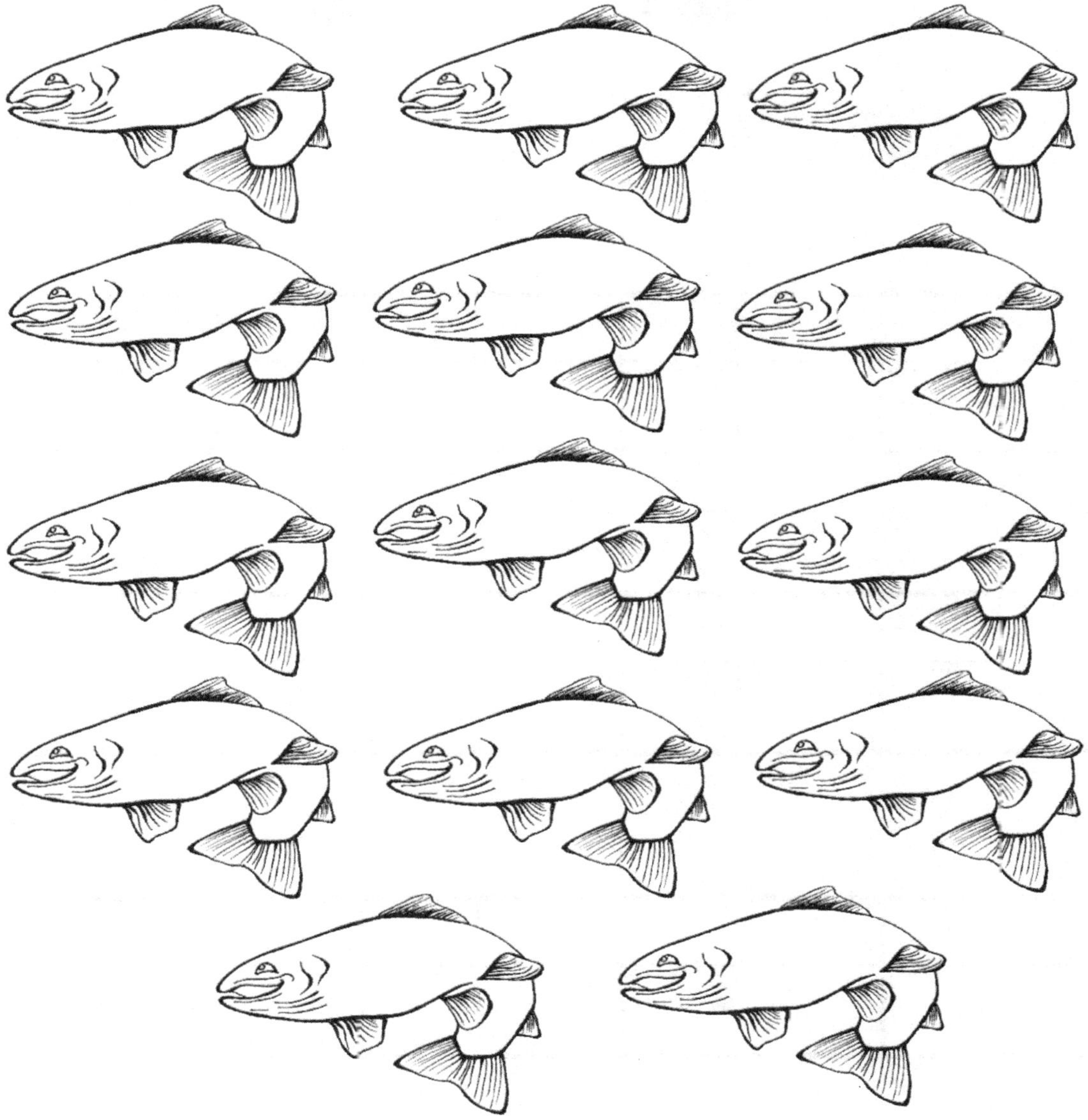

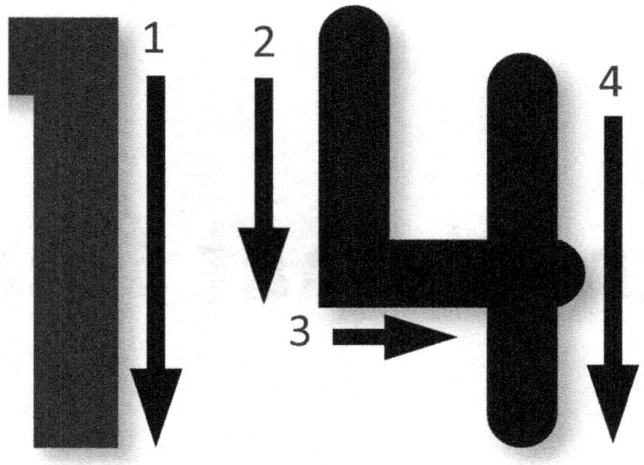
1
2
3
4

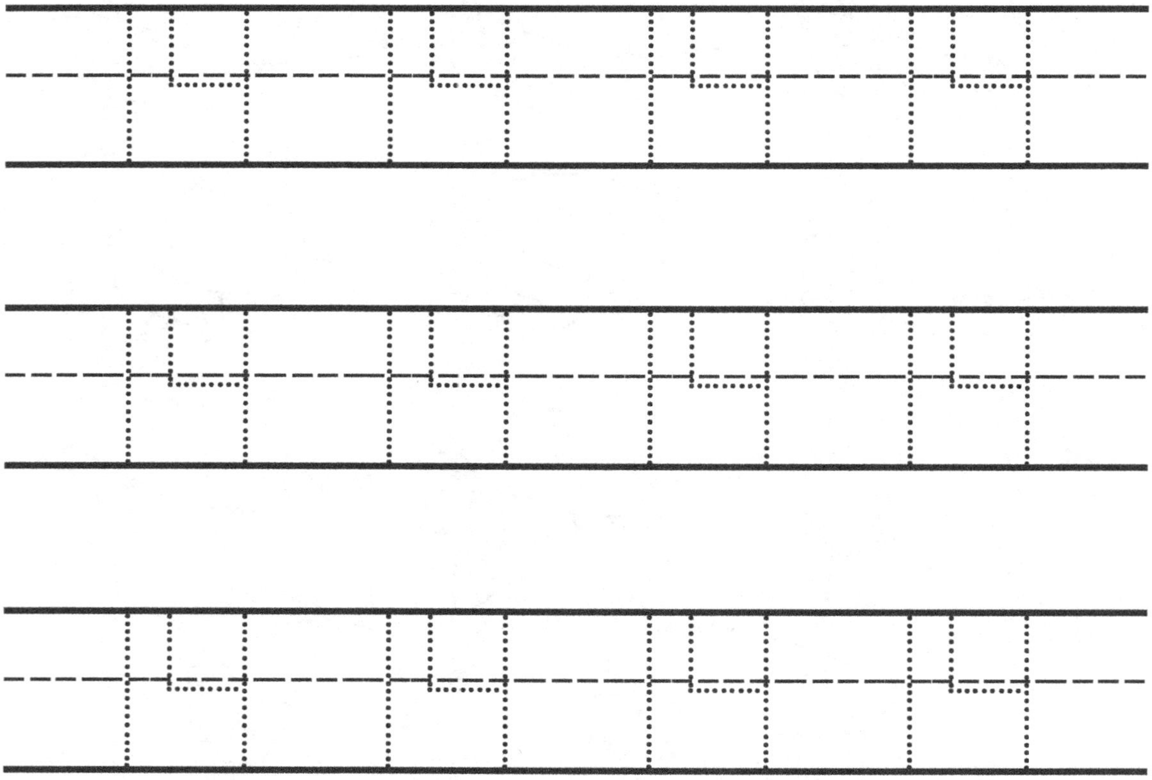

Thank You God for the

15 snails that I can count!

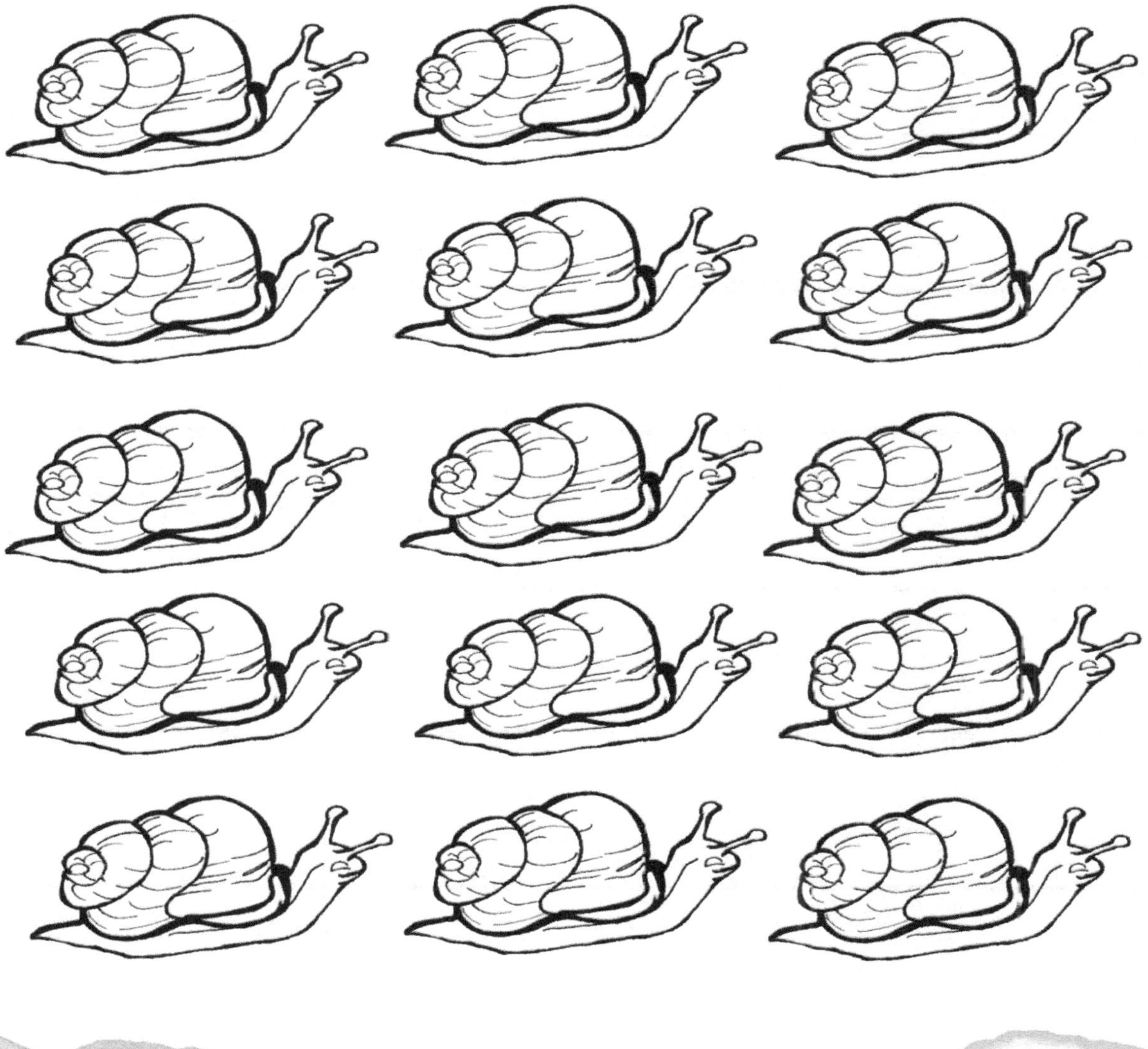

15 15 15 15

15 15 15 15

15 15 15 15

Thank You God for the

16 bees that I can count!

16 16 16 16

16 16 16 16

16 16 16 16

Thank You God for the

17 mushrooms that I can

count!

17
1
2
3

Thank You God for the

18 caterpillars that I can

count!

1
2
3
4
5

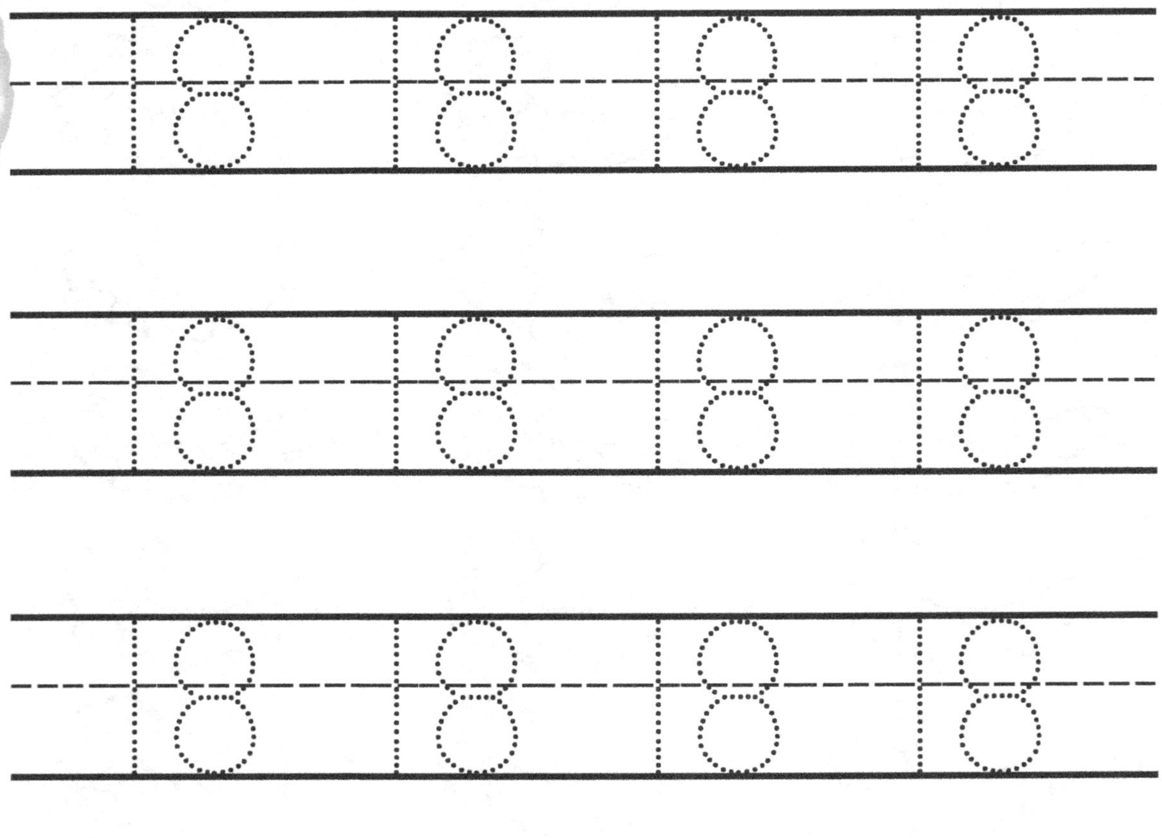

Thank You God for the

19 leaves that I can count!

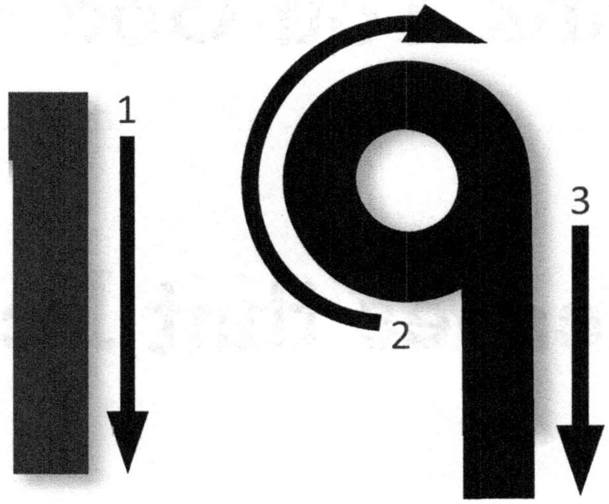
1
2
3

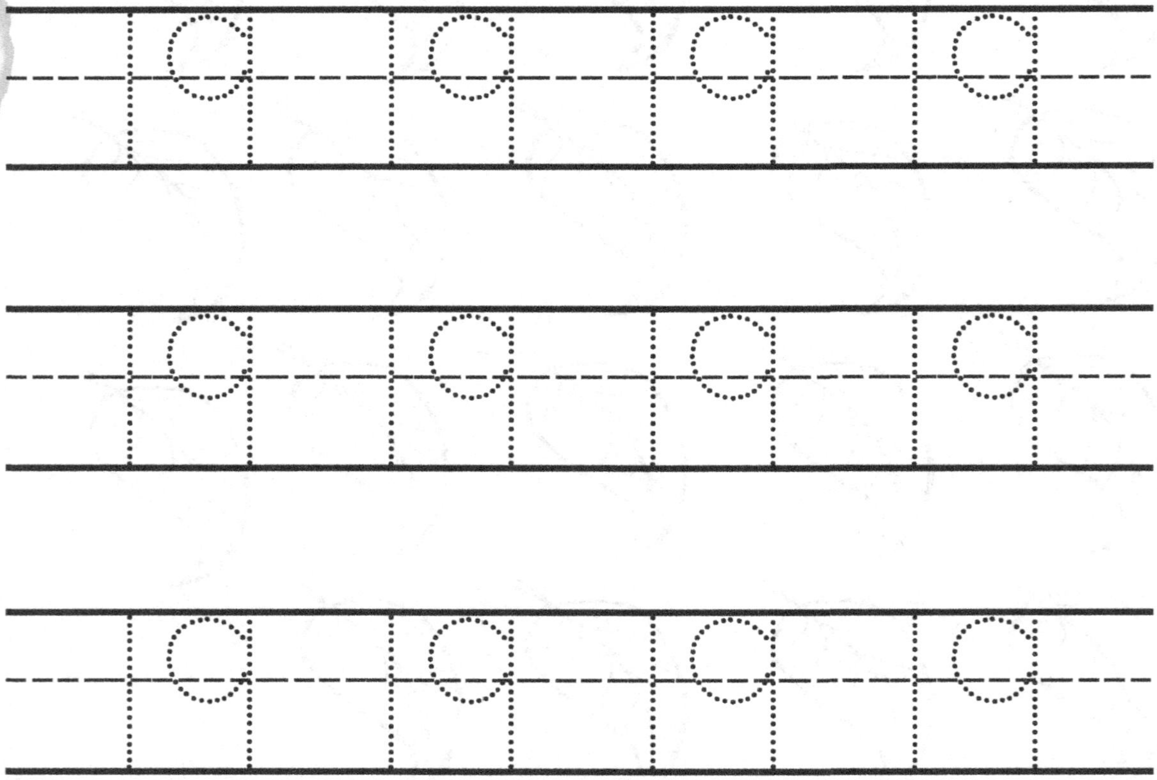

Thank You God for the

20 pumpkins that I can

count!

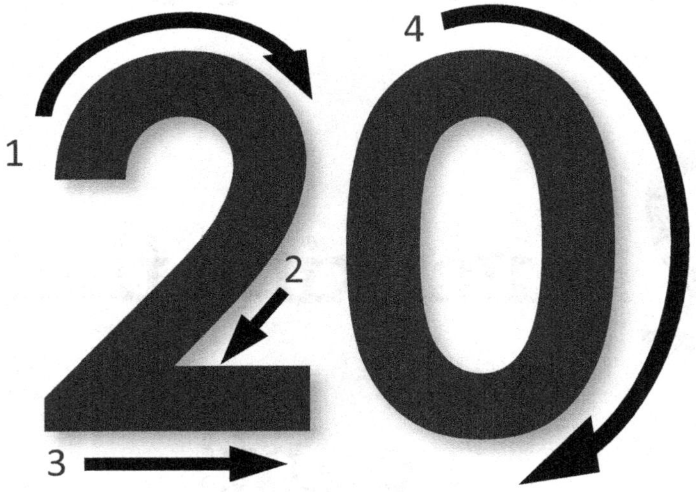

20 20 20

20 20 20

20 20 20

Section 2: Counting

We are planning a garden and you have a very important job. You need to count all of the plants and creatures we need to make fruit and vegetables grow. Let's get started!

Worms help make the fertile soil that plants need. How many worms can you count?

____

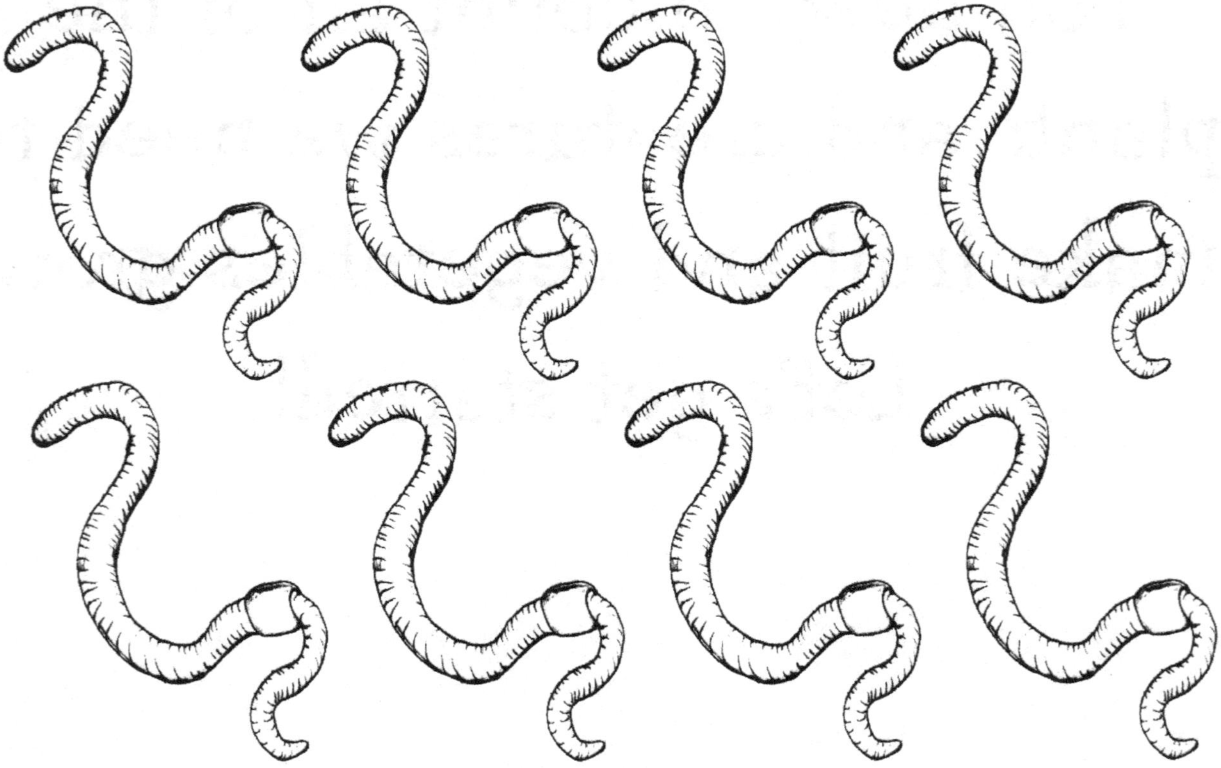

If flowers are pollinated, they can grow into fruits and vegetables, so we need you to count all of the flowers of the garden to find out how many fruits and vegetables could grow. Let's count!

How many zucchini flowers do you count?

How many watermelon flowers do you count?

How many strawberry flowers do you count?

____

How many apple blossoms do you count?

____

How many cherry blossoms do you count?

____

How many cucumber flowers do you count?

____

Thank You God for Blessing us with all of these beautiful flowers that can develop into tasty treats! And thank You for insects that pollinate plants by moving the pollen from one flower to another to make fruits and vegetables grow!

Let's count the pollinating insects! How many butterflies can you count?

____

How many beetles can you count?

____

How many ants can you count?

____

How many ladybug beetles can you count?

____

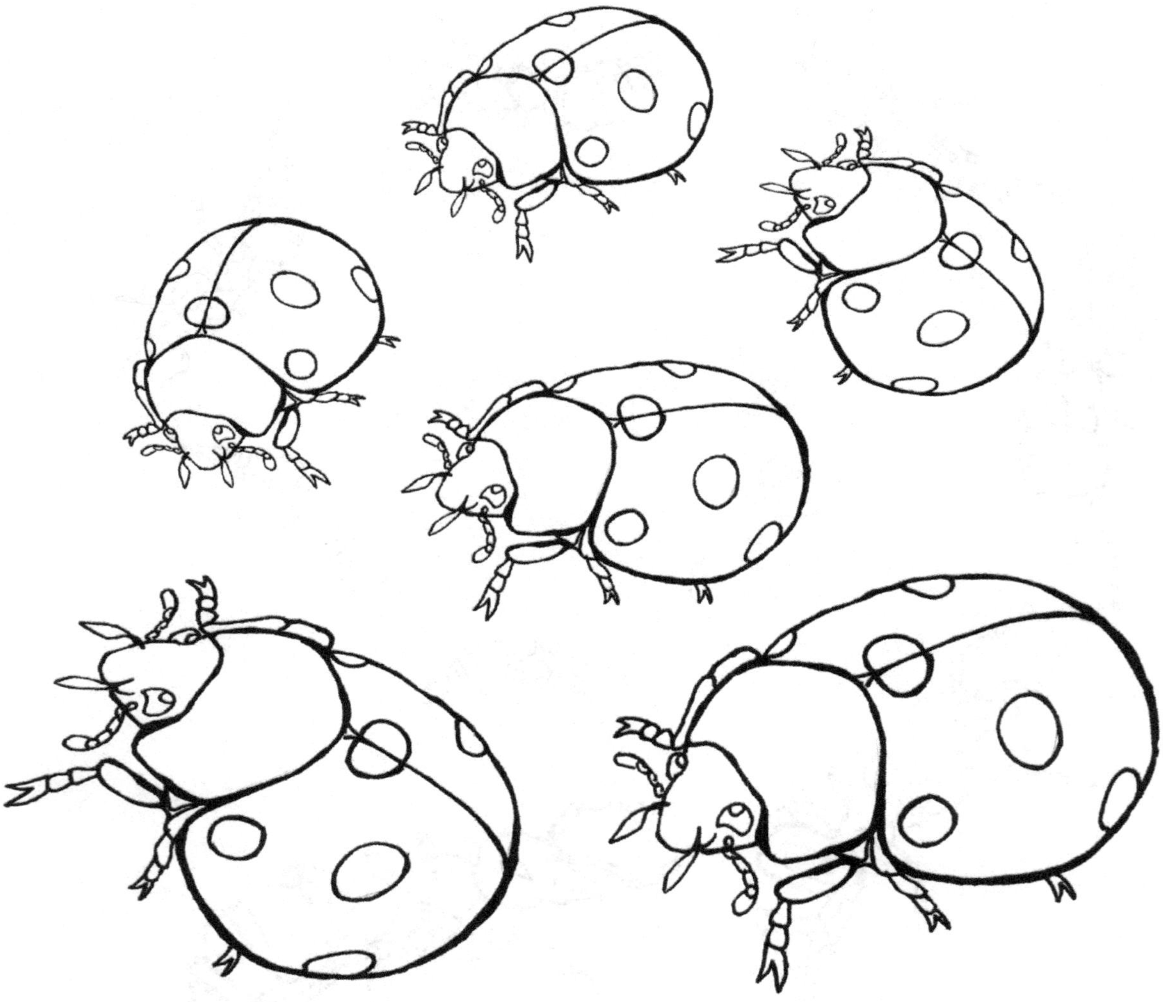

How many crickets can you count?

____

How many bees can you count?

____

Congratulations! You counted and made sure there were enough worms, flowers, and pollinators to make the fruits and veggies grow!

Section 2: Pattern Recognition

There are lots of patterns in nature, they are like hints telling us what will come next!

Flowers blooming followed by an insect pollinating them, followed by fruits or vegetables growing is an example of a pattern because it repeats itself again and again.

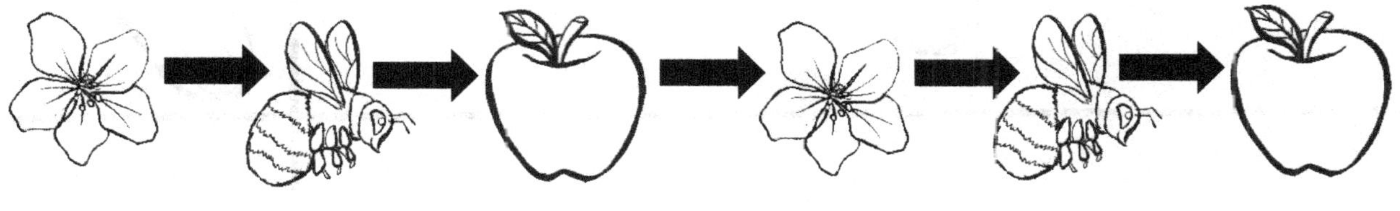

Can you figure out and circle the right thing to complete the pattern to make the garden grow?

Let's try more, circle the right thing to complete the pattern!

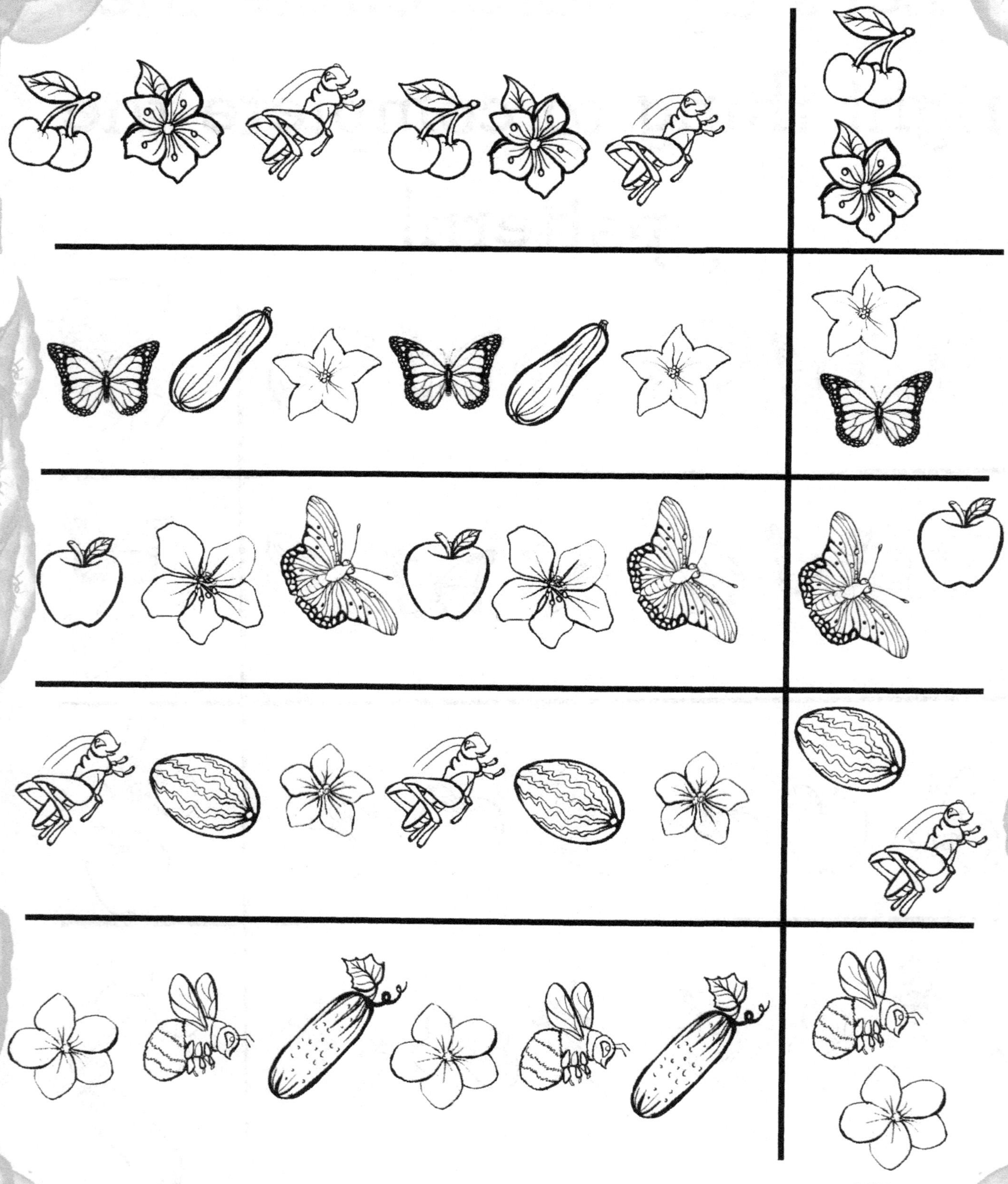

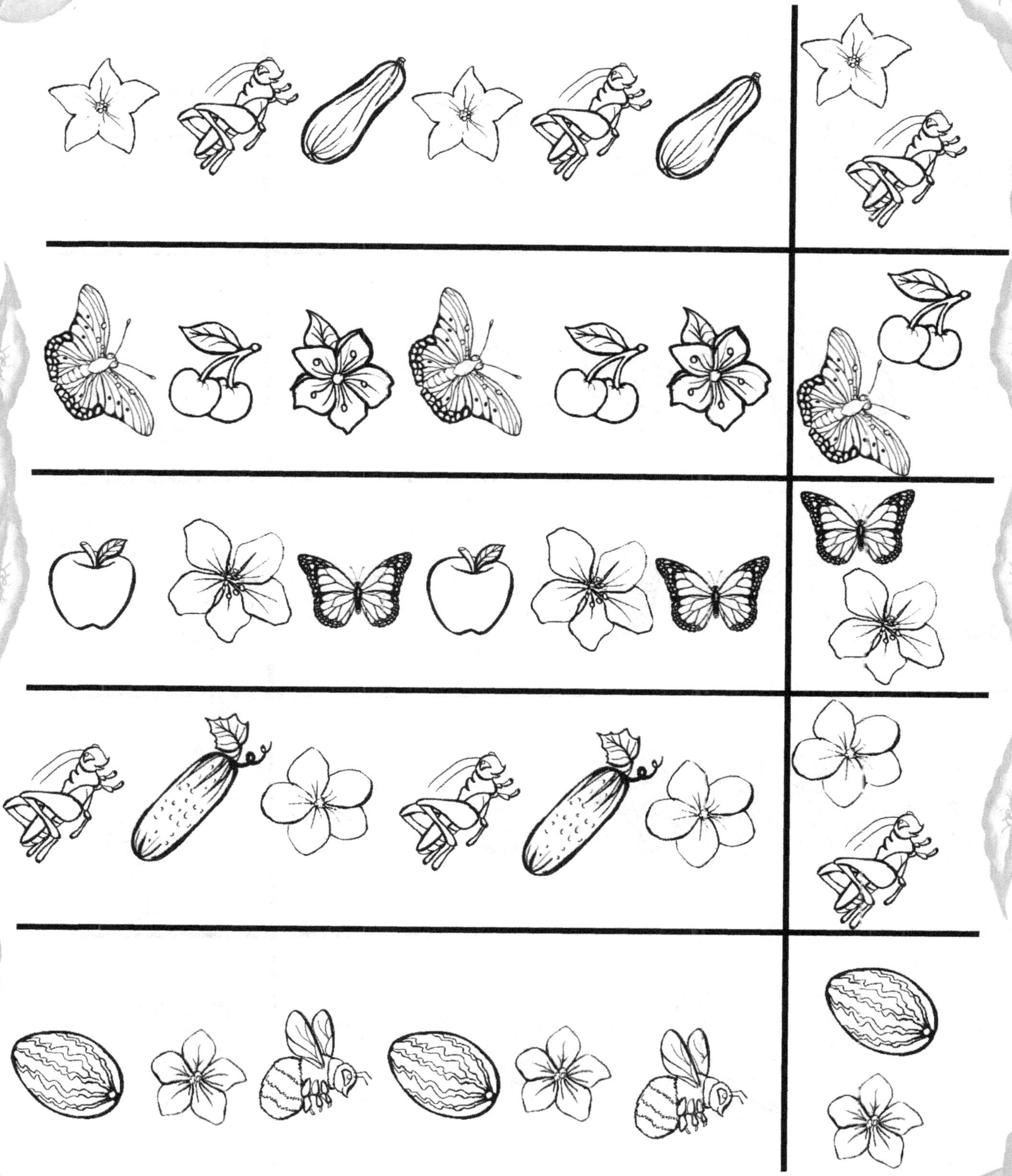

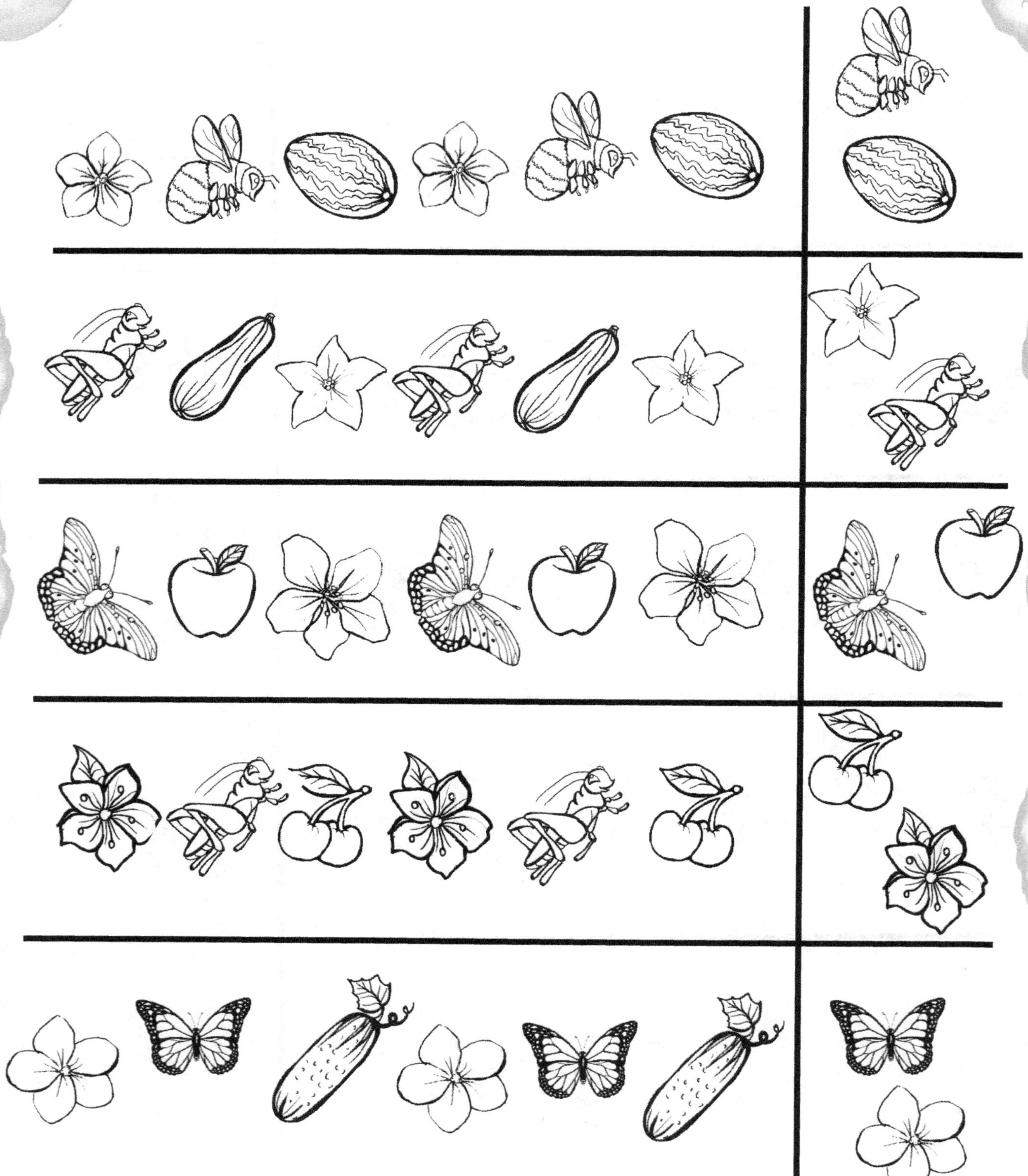

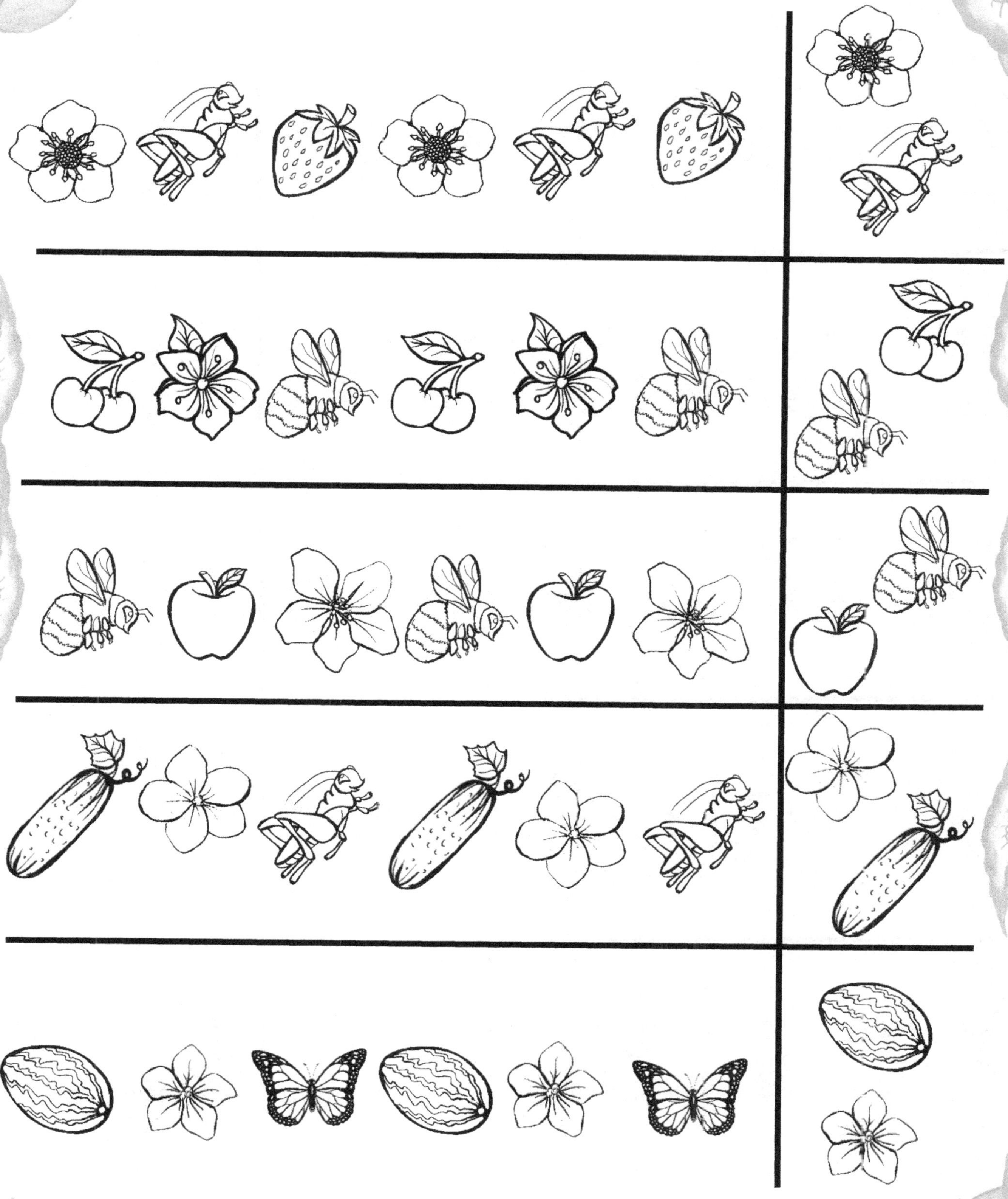

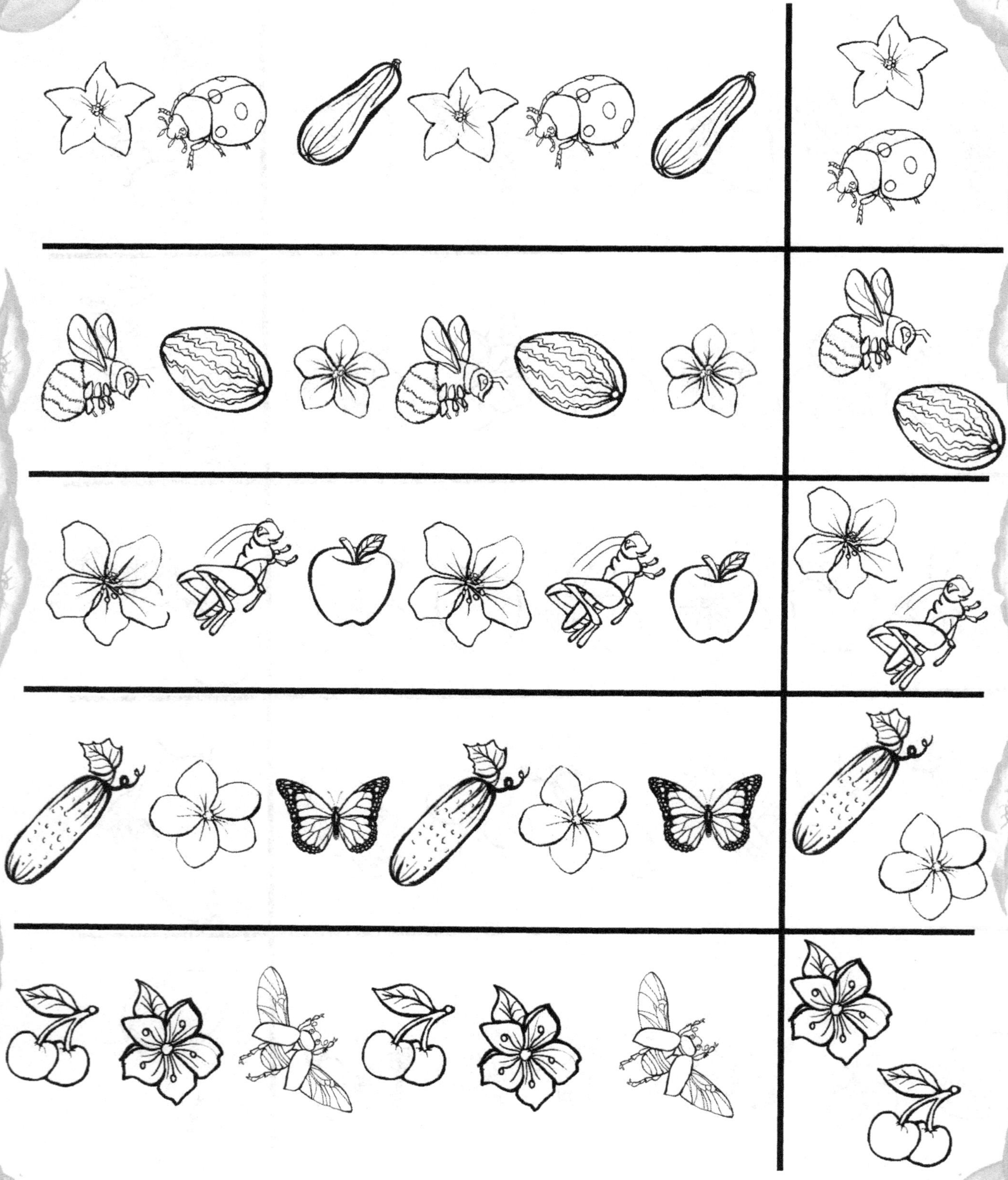

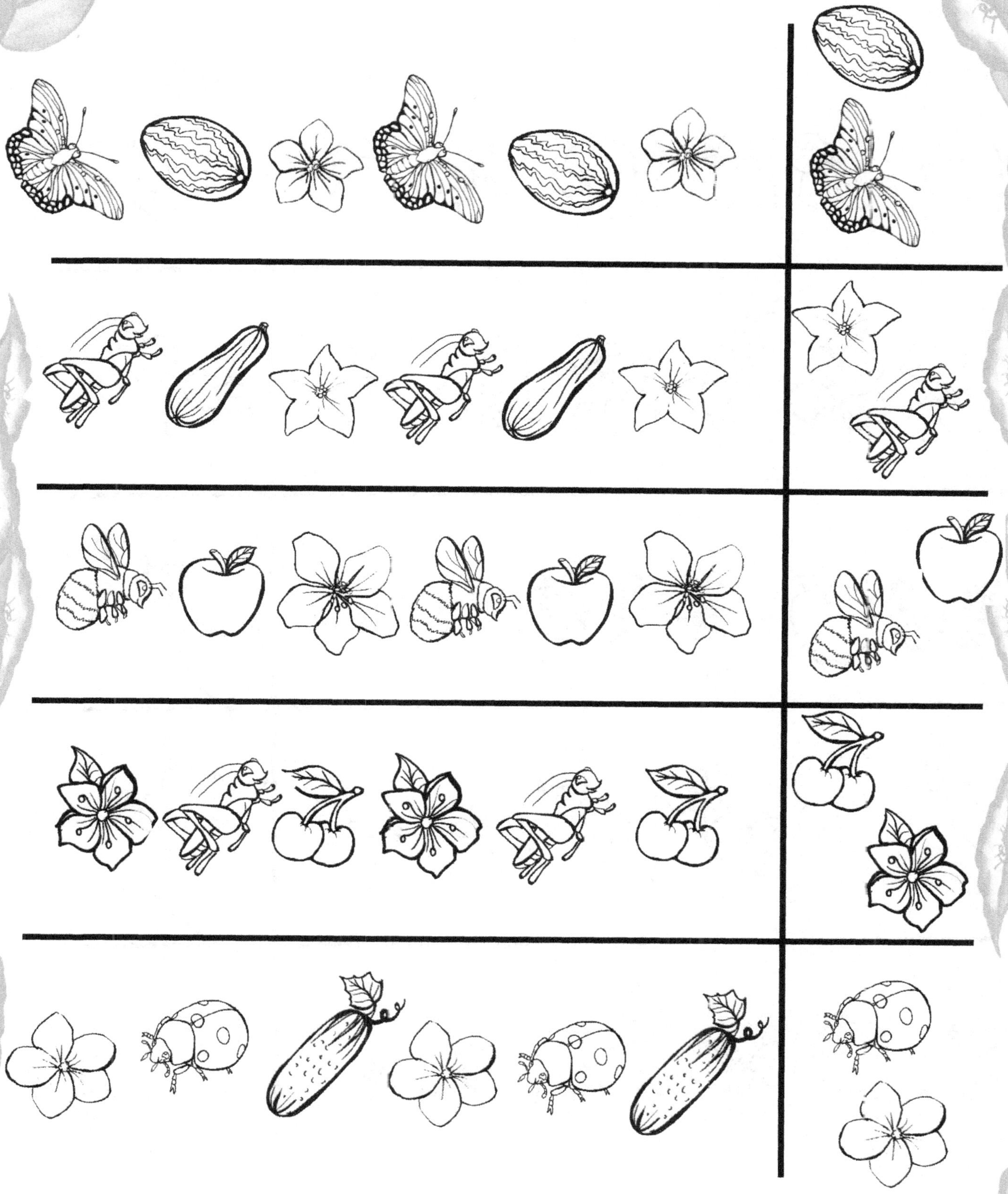

Thank You God for the pattern you created in nature of flowers, pollinators, and fruits and vegetables so that we can have delicious foods to eat!

We can also describe patterns we see in nature using numbers. For example, you can describe the dot pattern of this butterfly's abdomen as 1,2,1,2,1, and you can describe the dot pattern of the forewing as 2,4,2,4,2. Let's try and discover more patterns in nature!

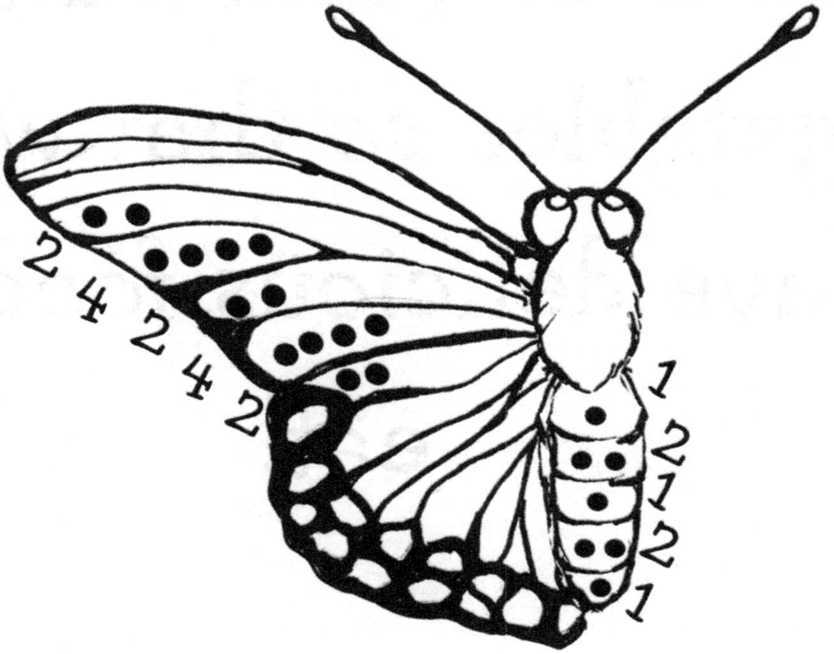

Can you count the dots on each segment of the caterpillars below and find the pattern?

___ ___ ___ ___ ___ ___ ___

___ ___ ___ ___ ___ ___ ___

___ ___ ___ ___ ___ ___ ___

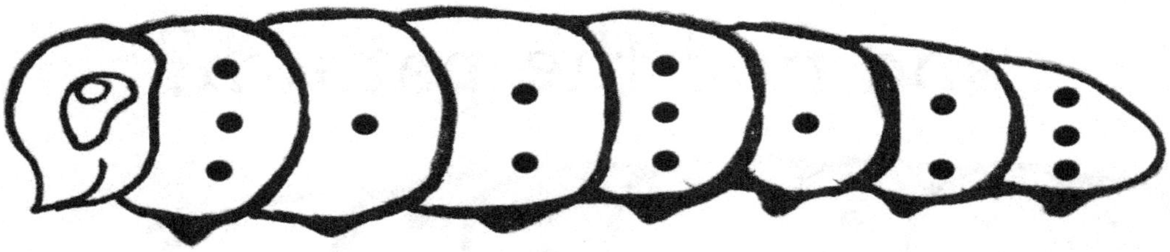

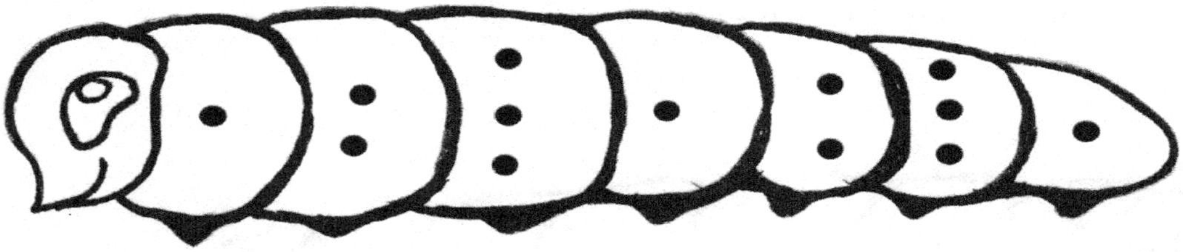

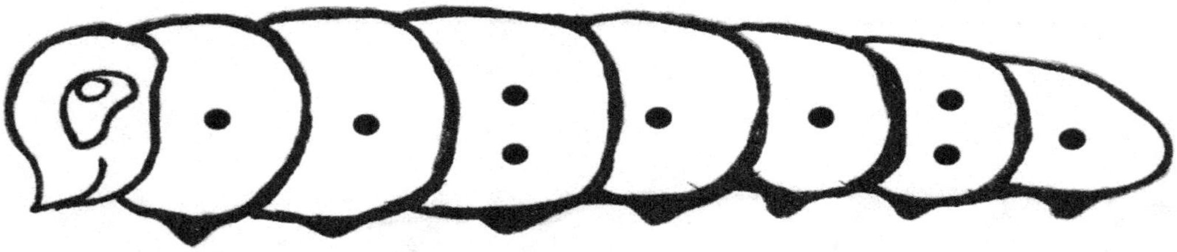

Patterns are also very important for growing a garden. Patterns help determine the correct spacing of plants that will ensure you can grow the most crops, but still have enough room for each to grow. Look at the pattern for growing a strawberry patch below.

2 1 2 1 2 1 2

Can you figure out the planting patterns below?

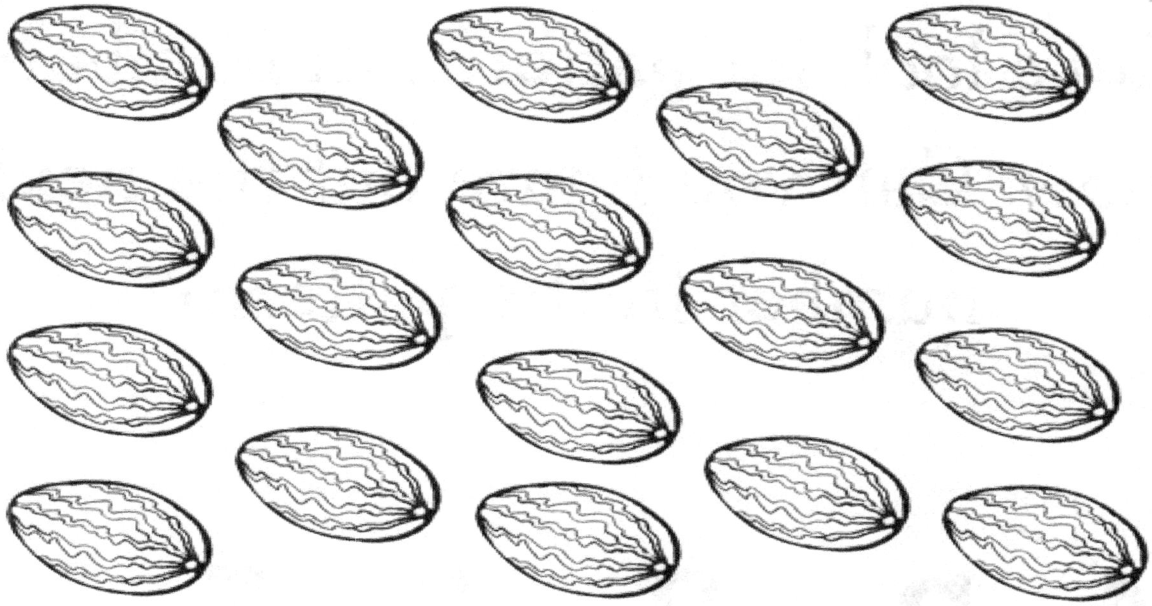

Let's look for patterns in numbers now. Can you figure out the next number in the pattern?

1 2 3 4 5 ___

2 4 6 8 10 ___

4 5 6 7 8 ___

0 3 6 9 12 ___

10 9 8 7 6 ___

7 6 5 4 3 ___

5 4 3 2 1 ___

9 8 7 6 5 ___

6 5 4 3 2 ___

8 7 6 5 4 ___

1 2 1 2 1 2 __

1 2 3 1 2 3 __

3 2 1 3 2 1 __

1 3 2 1 3 2 __

1 1 2 2 3 3 __

4 3 3 2 2 1 __

6 9 12 15 ____

12 10 8 6 ____

15 12 9 6 ____

16 12 10 8 ____

9 11 13 15 ____

0 5 10 15 ____

Thank You God for all of these interesting patterns and for blessing me with a mind to discover them!

Section 3: Addition & Subtraction

Subtraction and Addition happen all the time in nature, like trees as they lose leaves in fall and then gain leaves in spring.

Every autumn the leaves change color and fall from many kinds of trees. A loss of something, like leaves, is known as **subtraction**

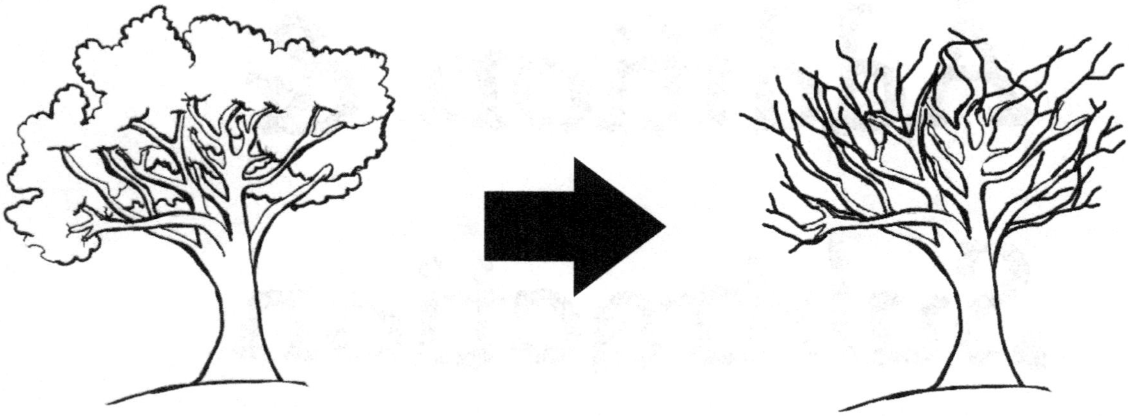

The leaves then come back in the spring. An increase in something, like leaves, is known as **addition.**

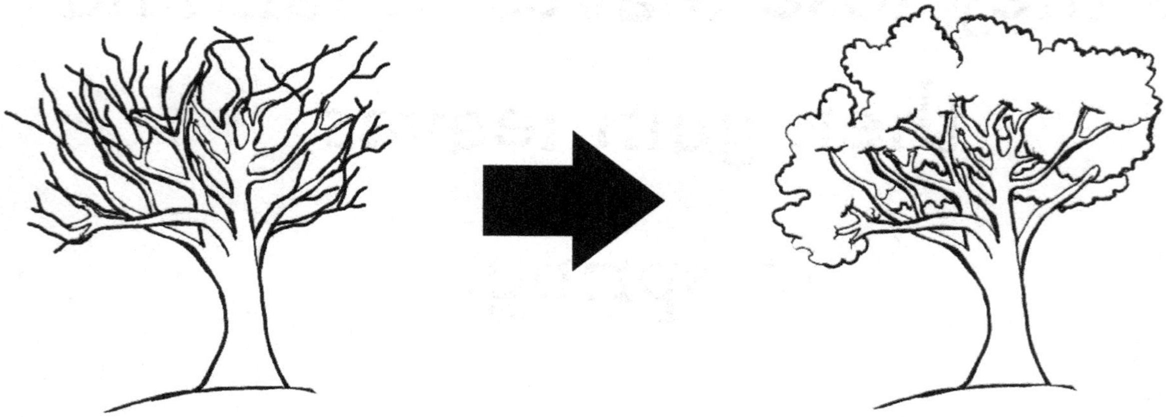

Let's find the increase of leaves in the spring by adding the leaves as they grow!

 +  = 2

 +  =

 +  =

 +  =

 +  =

 +  =

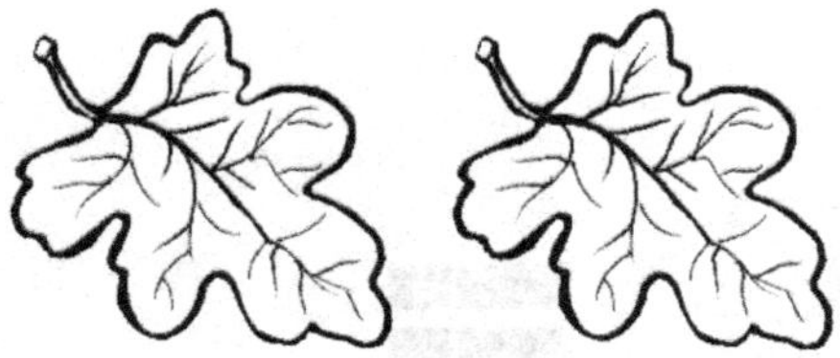 +  =

 + 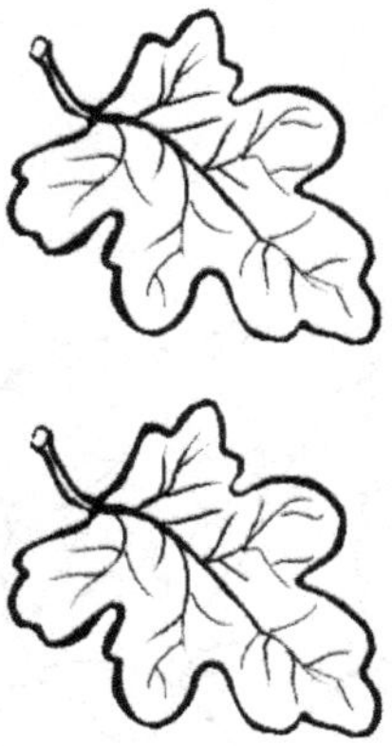 =

 +  =

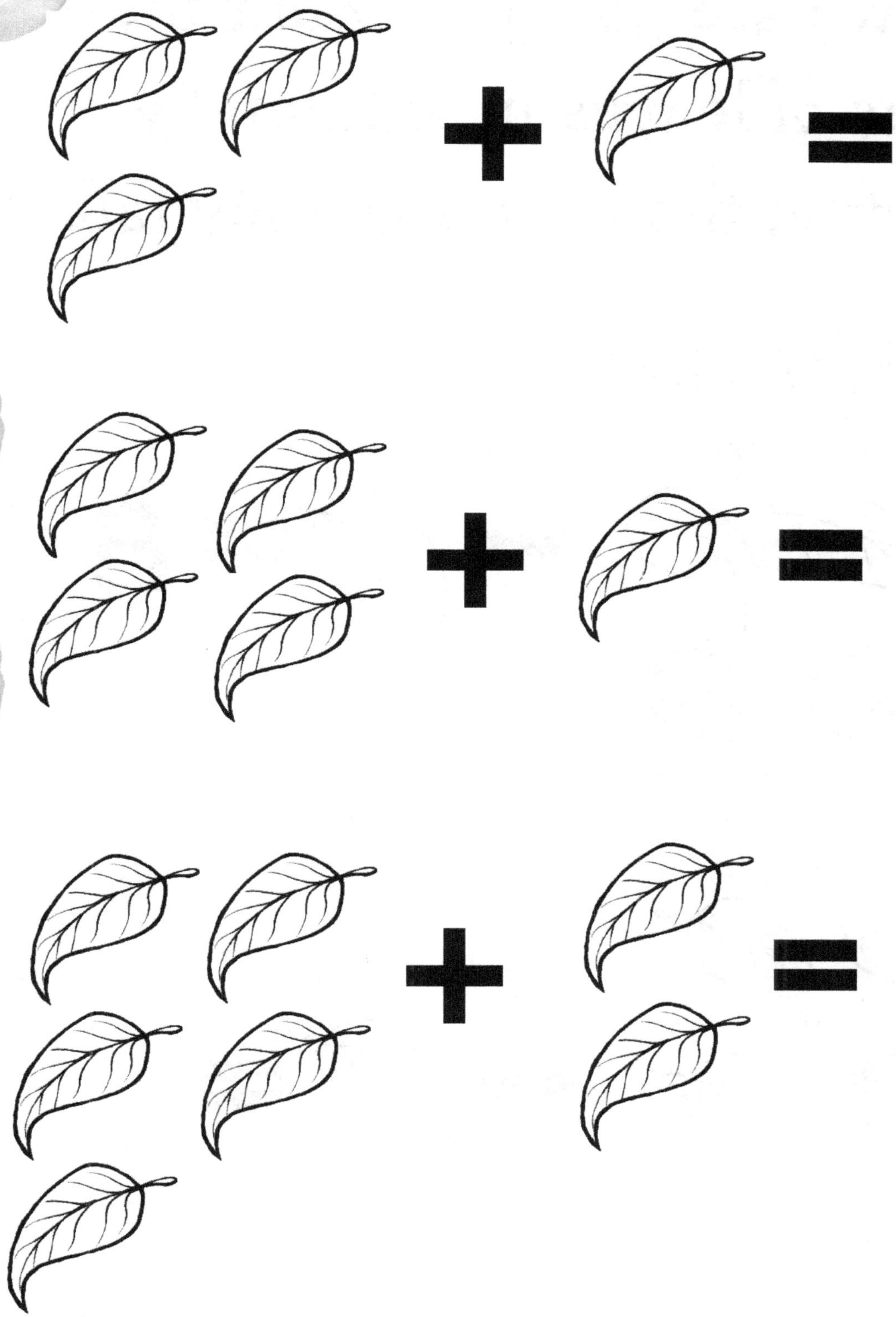

Now we are going to count the losses of leaves in autumn by subtracting the leaves as they fall.

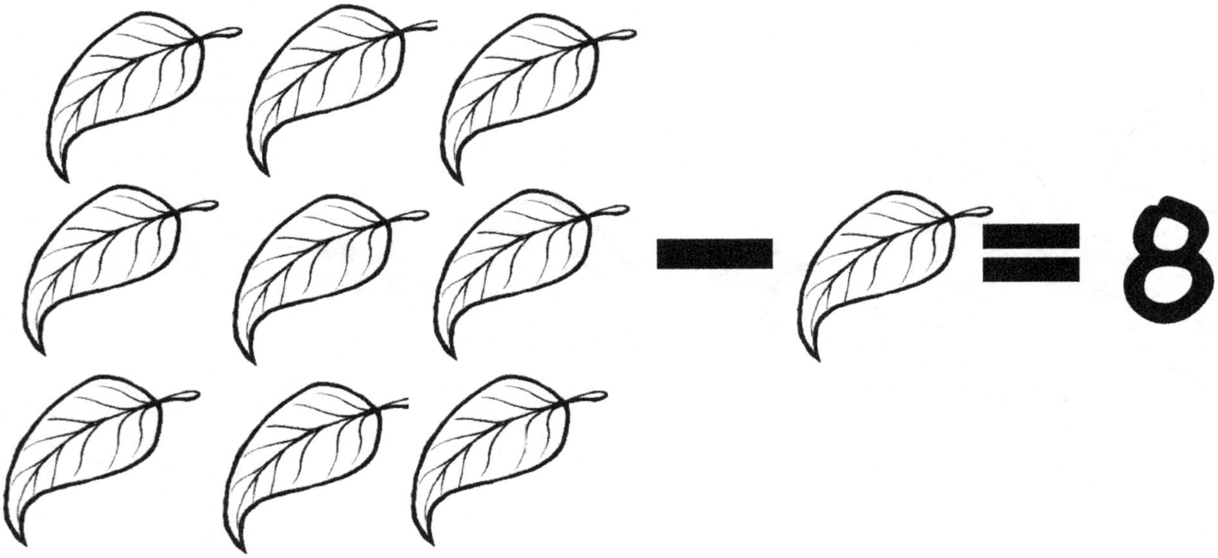

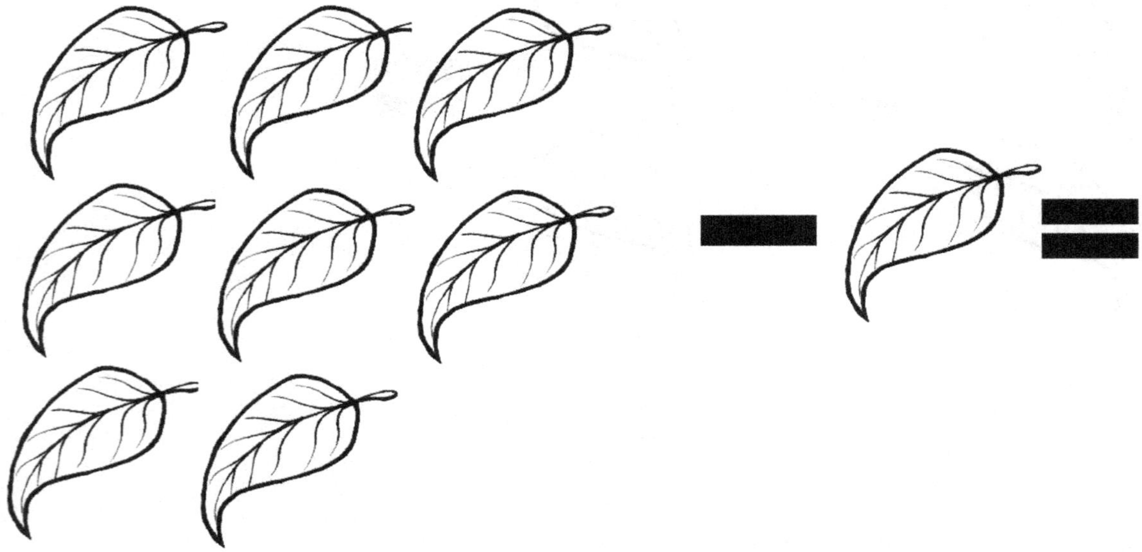

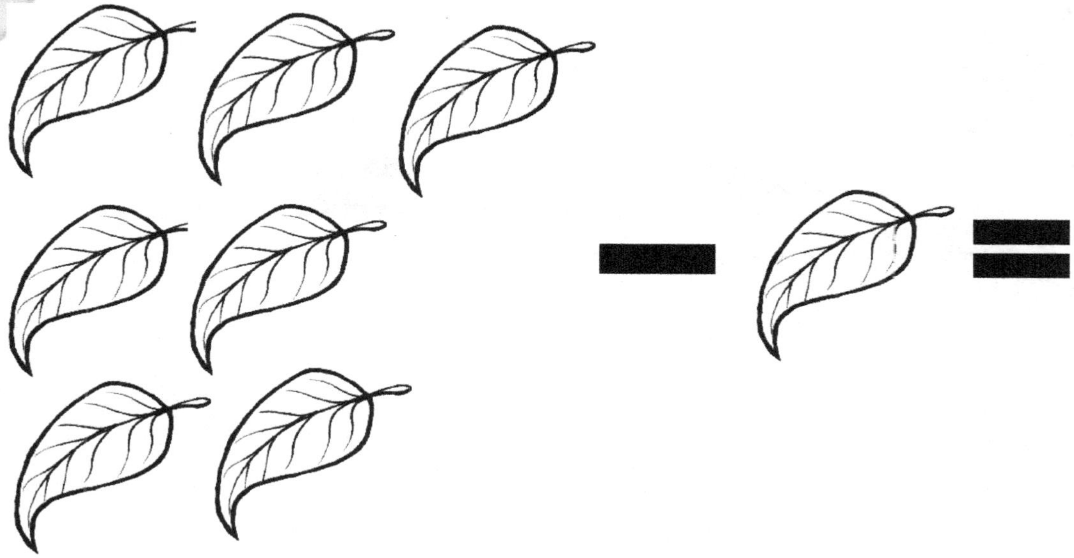

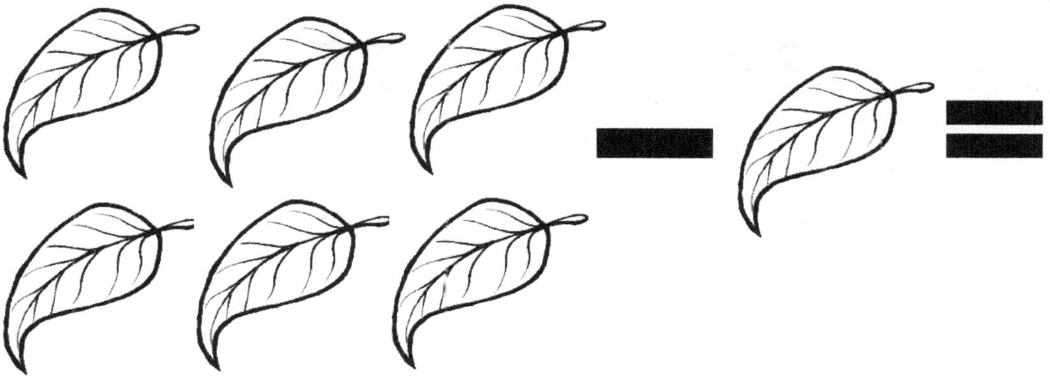

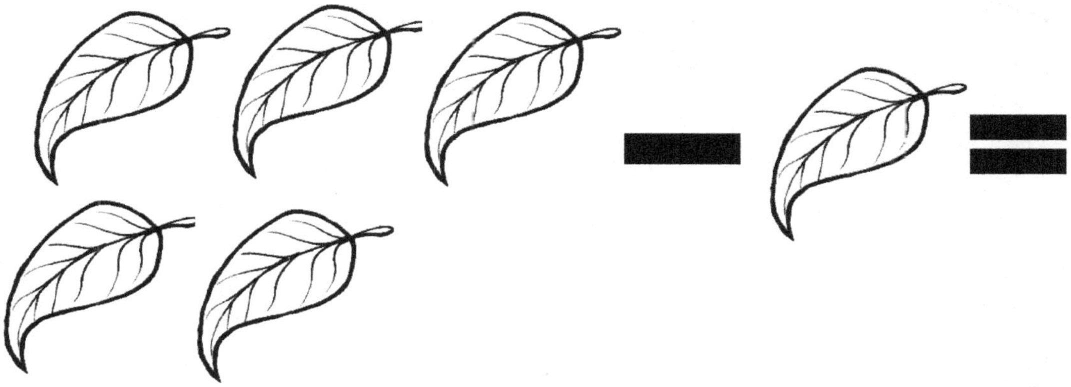

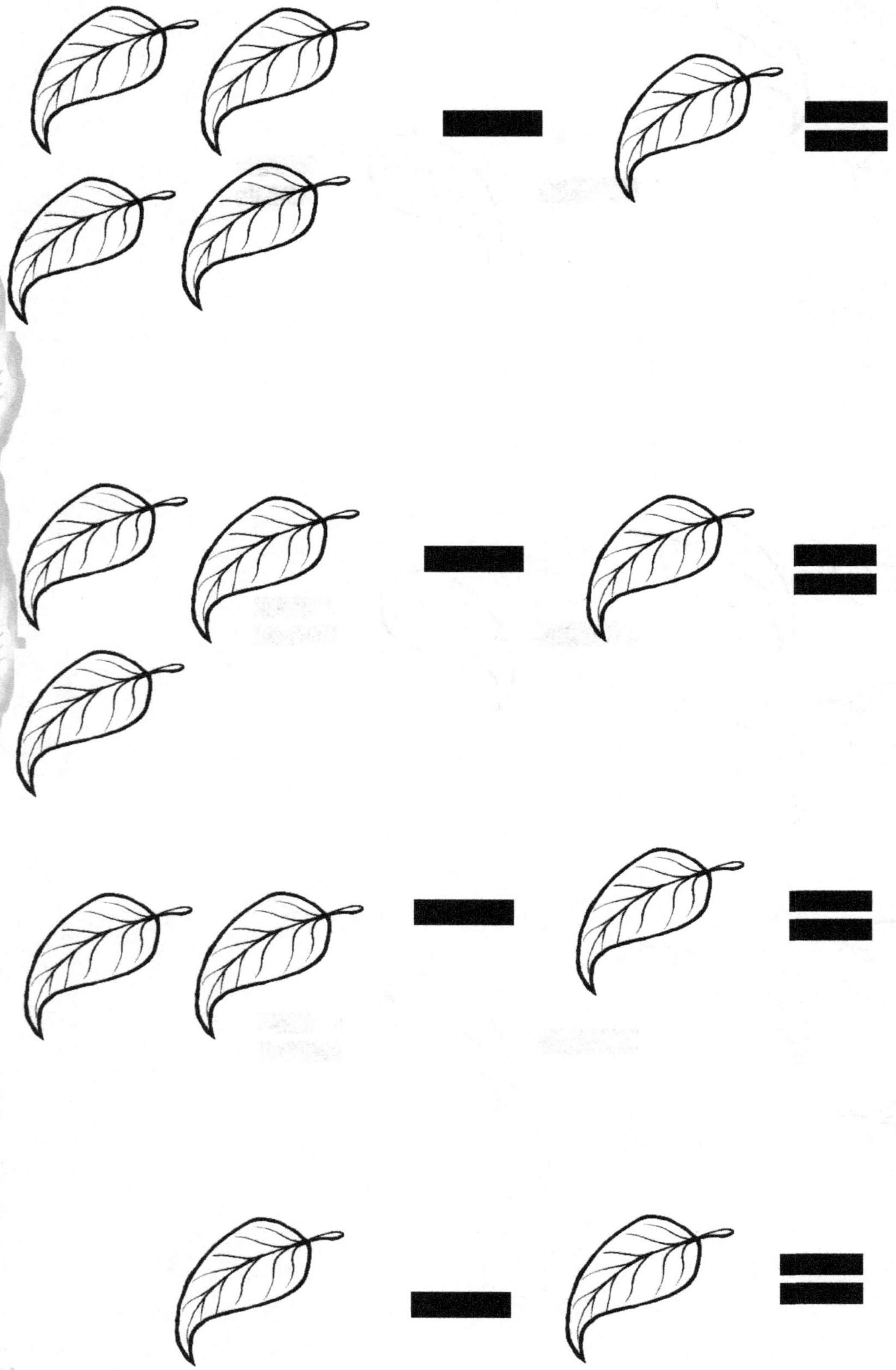

 − 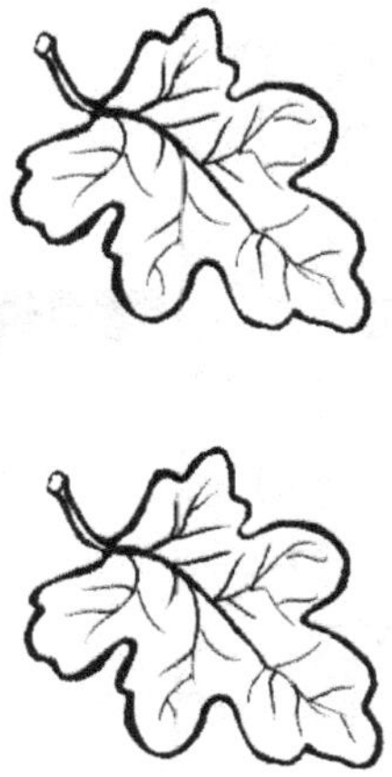 =

 −  =

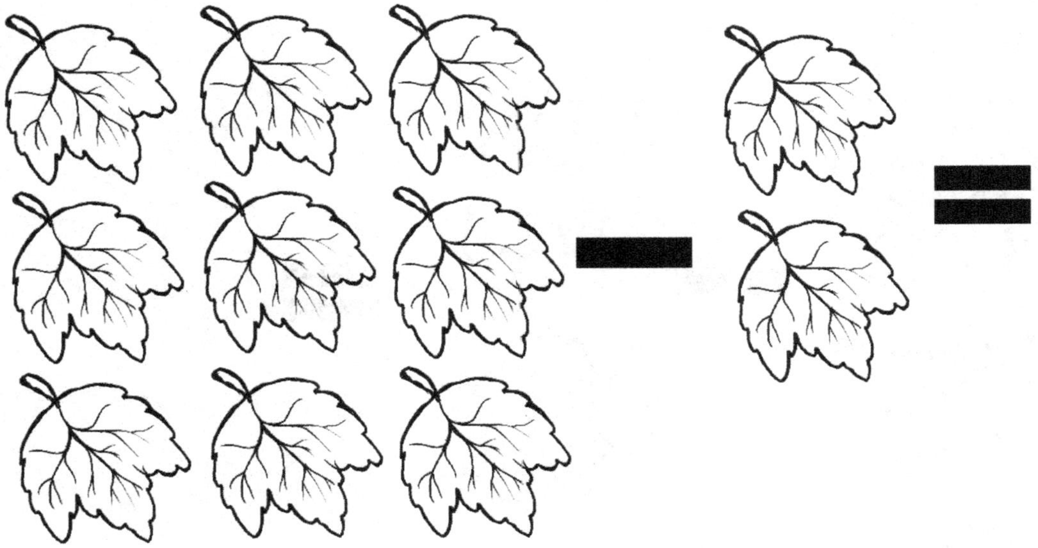

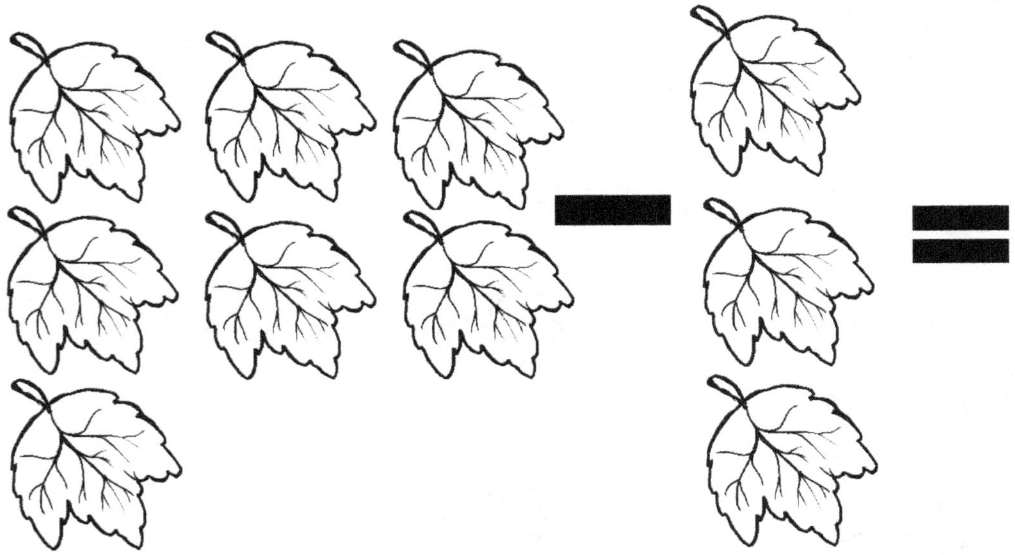

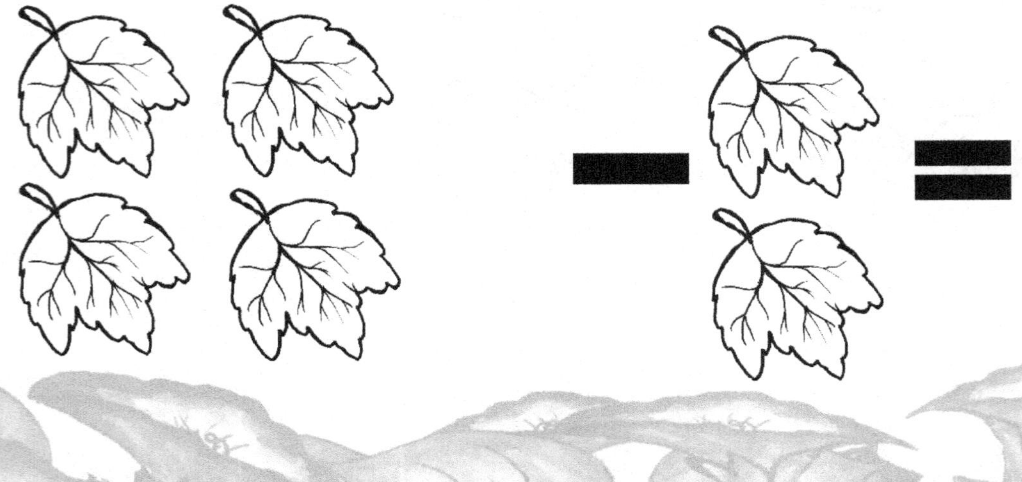

Every time a new flower grows in the garden that is addition. Let's practice addition by watching the flowers grow!

First there are **14** flowers and then **6** more start growing.

Now there are **20** flowers growing.

Let's try more, add the flowers to find out how many flowers there are and right down your answer.

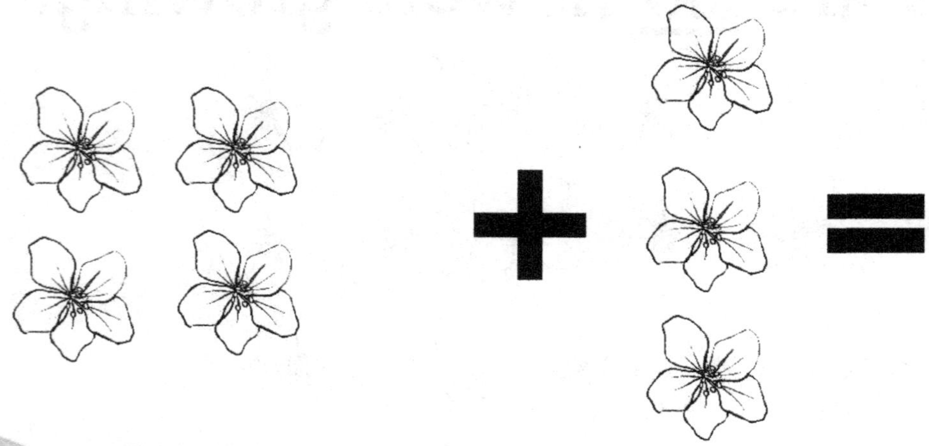

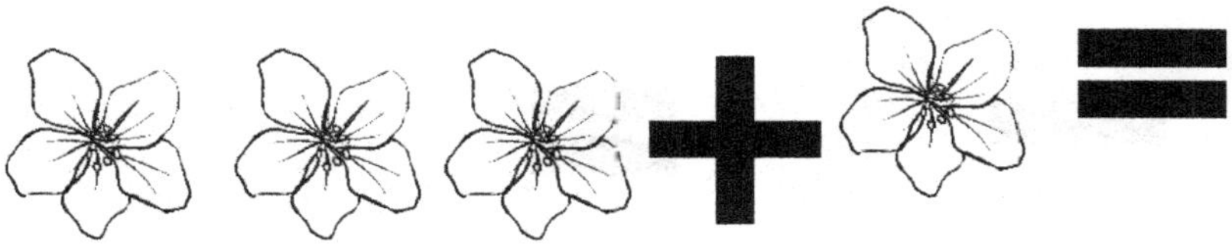

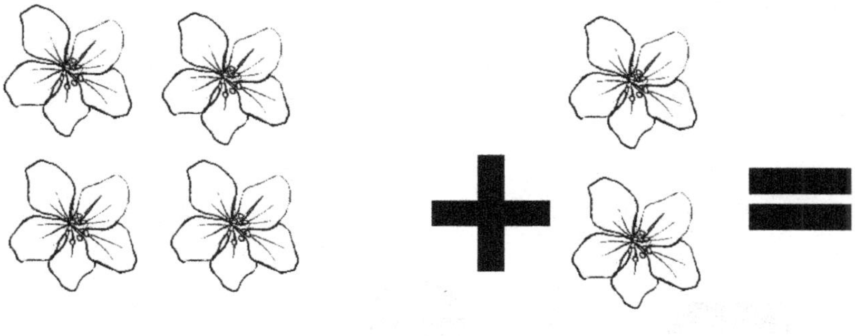

Flowers only bloom for a certain time, some are eaten by animals and some die naturally. We are going to keep track of our garden by counting the loss of flowers through using subtraction.

 − 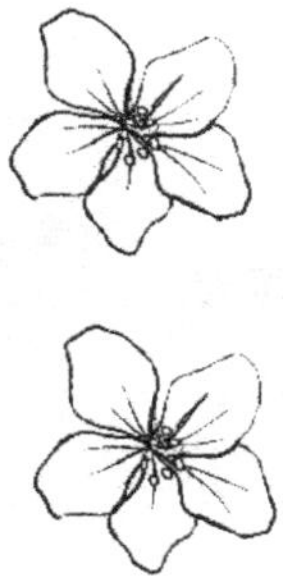 =

 −  =

 −  =

 − 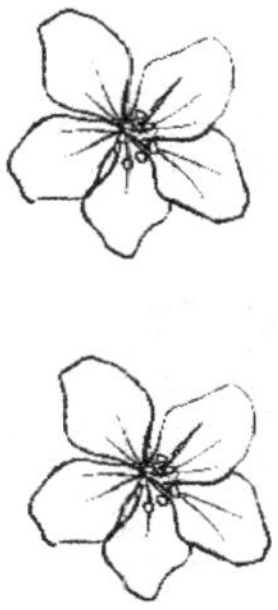 =

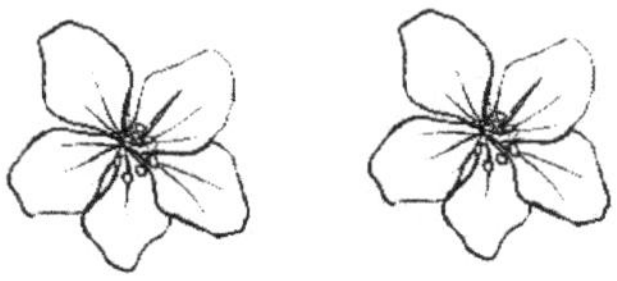 −  =

 −  =

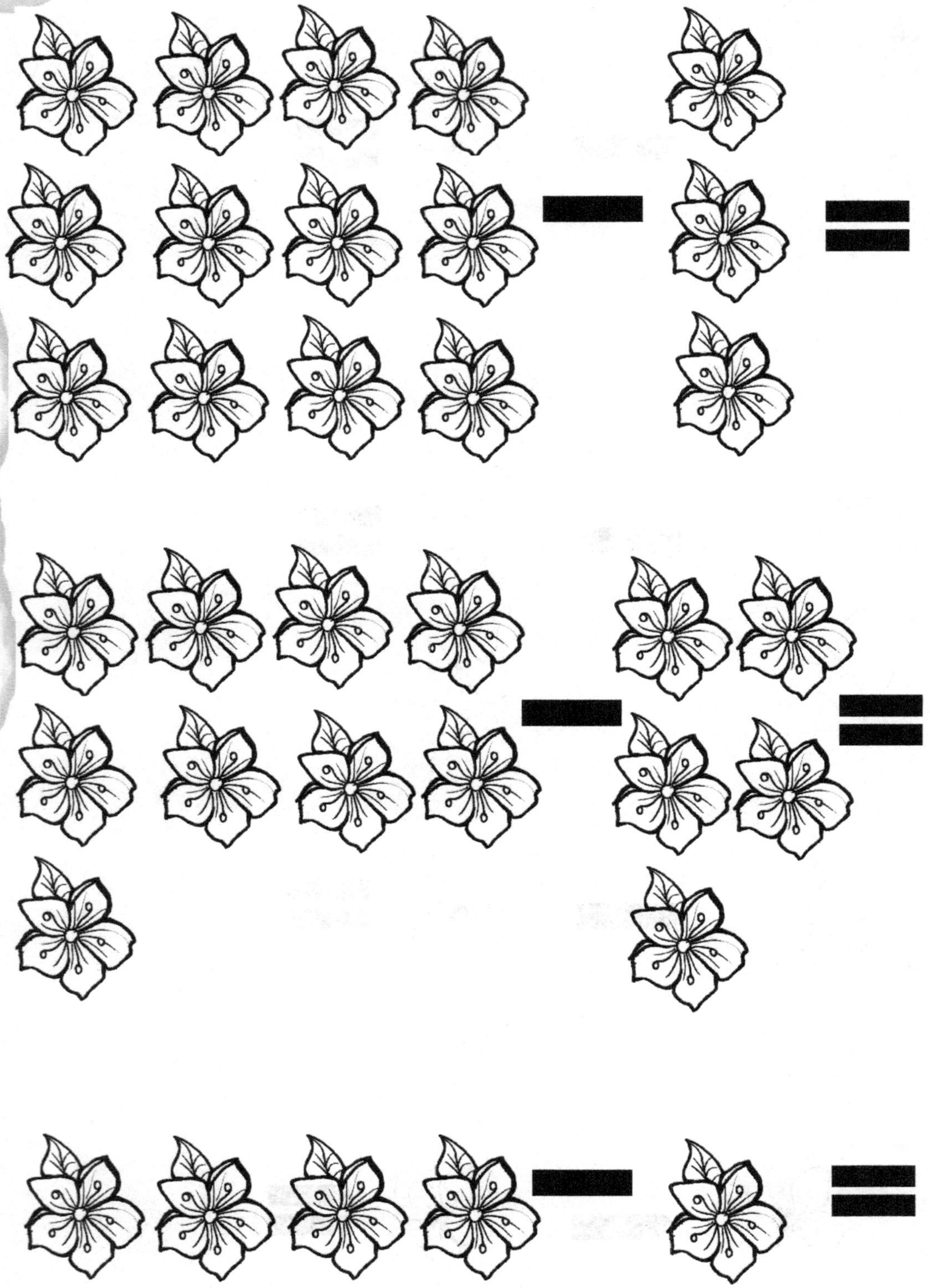

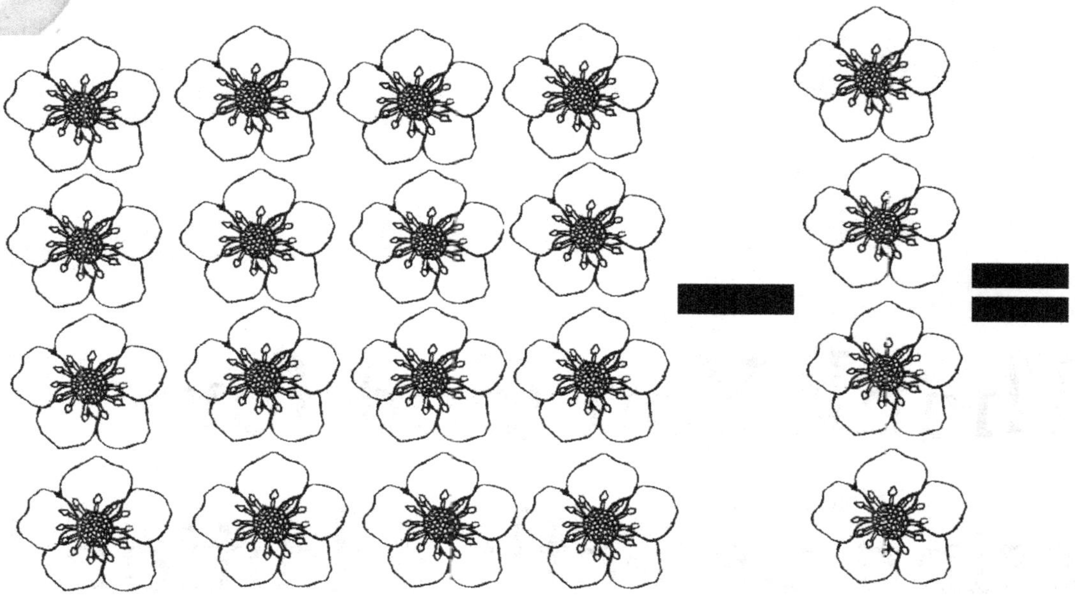

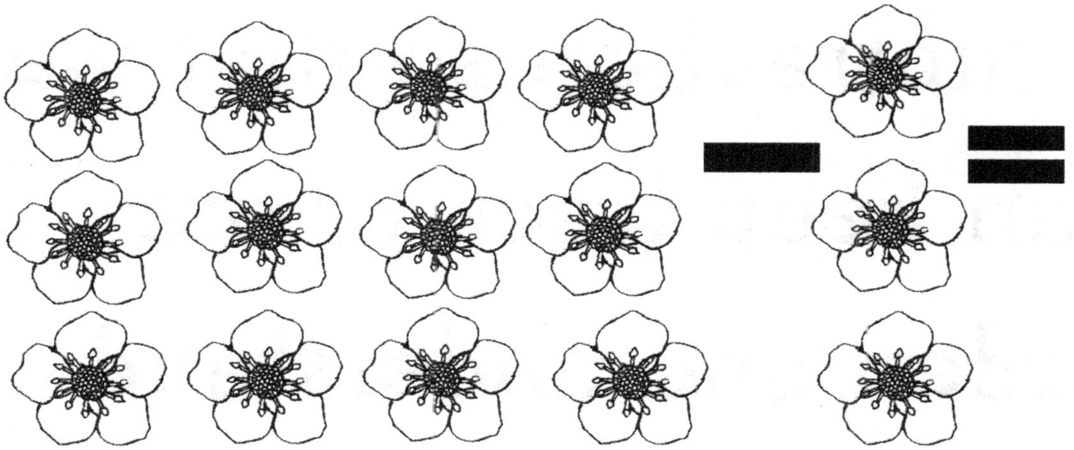

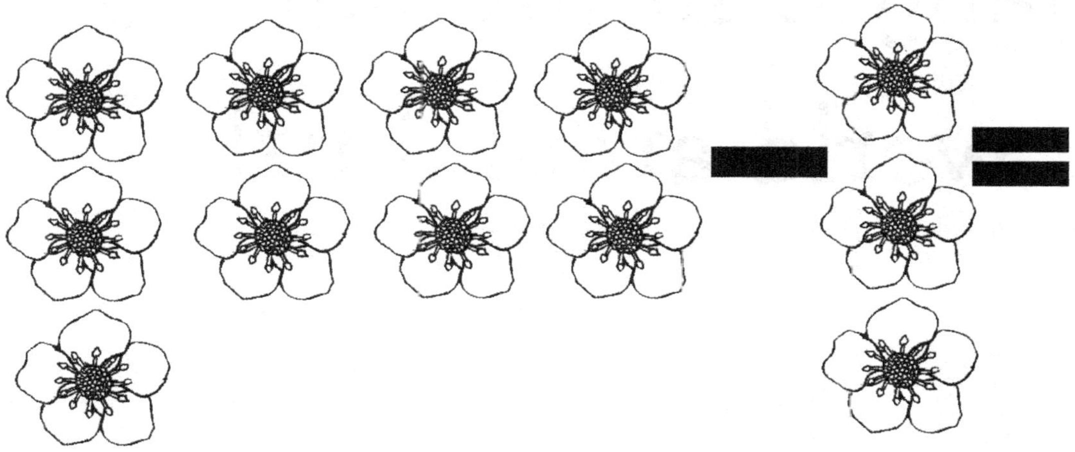

Thank You God for blessing us with addition and subtraction so that we can keep track of our garden and understand changes in the natural world around us!

Let's practice more addition! Can you add the numbers below?

1 + 2 = ___

3 + 2 = ___

5 + 3 = ___

7 + 3 = ___

2 + 9 = ___

1 + 6 = ___

4 + 3 = ___

2 + 2 = ___

2 + 4 = ___

6 + 4 = ___

8 + 1 = ___

2 + 9 = ___

7 + 7 = ___

6 + 5 = ___

2 + 8 = ___

1 + 5 = ___

5 + 5 = ___

$6 + 7 = \_\_\_$

$7 + 8 = \_\_\_$

$9 + 3 = \_\_\_$

$5 + 9 = \_\_\_$

$8 + 9 = \_\_\_$

$8 + 8 = \_\_\_$

$2 + 8 = \_\_\_$

$4 + 7 = \_\_\_$

$3 + 7 = \_\_\_$

$6 + 6 = \_\_\_$

$6 + 1 = \_\_\_$

$3 + 3 = \_\_\_$

$3 + 4 =$ ____

$1 + 4 =$ ____

$5 + 1 =$ ____

$4 + 2 =$ ____

$2 + 7 =$ ____

$1 + 9 =$ ____

$9 + 9 =$ ___

$4 + 6 =$ ___

$1 + 7 =$ ___

$4 + 4 =$ ___

$3 + 3 =$ ___

$3 + 6 =$ ___

0 + 1 = ___

1 + 0 = ___

0 + 2 = ___

2 + 0 = ___

0 + 3 = ___

3 + 0 = ___

0 + 4 = ___

0 + 5 = ___

0 + 6 = ___

7 + 0 = ___

8 + 0 = ___

9 + 0 = ___

13 + 5 = ___

14 + 6 = ___

18 + 1 = ___

17 + 3 = ___

10 + 2 = ___

12 + 3 = ___

Let's practice more subtraction! Can you subtract the second number from the first number?

$2 - 1 =$ __

$4 - 2 =$ __

$5 - 1 =$ __

$7 - 3 =$ __

$9 - 5 =$ __

$3 - 1 = \_\_\_$

$9 - 2 = \_\_\_$

$8 - 3 = \_\_\_$

$9 - 3 = \_\_\_$

$7 - 2 = \_\_\_$

$5 - 4 = \_\_\_$

$8 - 1 =$ ___

$8 - 4 =$ ___

$7 - 4 =$ ___

$3 - 2 =$ ___

$4 - 3 =$ ___

$6 - 1 =$ ___

9 – 8 = ___

6 – 2 = ___

7 – 3 = ___

6 – 5 = ___

5 – 2 = ___

7 – 1 = ___

$5 - 3 = \_\_\_$

$7 - 6 = \_\_\_$

$9 - 1 = \_\_\_$

$9 - 6 = \_\_\_$

$5 - 3 = \_\_\_$

$4 - 1 = \_\_\_$

$1 - 1 = \_\_\_$

$7 - 7 = \_\_\_$

$8 - 8 = \_\_\_$

$9 - 9 = \_\_\_$

$4 - 4 = \_\_\_$

$5 - 5 = \_\_\_$

$1 - 0 = \_\_\_$

$2 - 0 = \_\_\_$

$3 - 0 = \_\_\_$

$4 - 0 = \_\_\_$

$5 - 0 = \_\_\_$

$6 - 0 = \_\_\_$

10 – 0 = ____

12 – 0 = ____

18 – 0 = ____

14 – 0 = ____

16 – 0 = ____

19 – 0 = ____

19 – 2 = ___

20 – 3 = ___

13 – 7 = ___

16 – 4 = ___

12 – 7 = ___

15 – 5 = ___

20 – 10 = ___

13 – 12 = ___

15 – 14 = ___

19 – 18 = ___

17 – 13 = ___

13 – 10 = ___

7 – 7 = ___

8 – 8 = ___

6 – 6 = ___

3 – 3 = ___

2 – 2 = ___

4 – 4 = ___

$12 - 12 = \underline{\qquad}$

$15 - 15 = \underline{\qquad}$

$18 - 18 = \underline{\qquad}$

$10 - 10 = \underline{\qquad}$

$17 - 17 = \underline{\qquad}$

$16 - 16 = \underline{\qquad}$

$11 - 2 =$ ____

$11 - 0 =$ ____

$11 - 11 =$ ____

$8 - 8 =$ ____

$11 - 9 =$ ____

$5 - 4 =$ ____

$5 - 0 = \underline{\qquad}$

$6 - 2 = \underline{\qquad}$

$3 - 3 = \underline{\qquad}$

$6 - 2 = \underline{\qquad}$

$9 - 7 = \underline{\qquad}$

$9 - 9 = \underline{\qquad}$

$10 - 0 = \_\_\_$

$12 - 2 = \_\_\_$

$14 - 3 = \_\_\_$

$18 - 2 = \_\_\_$

$15 - 7 = \_\_\_$

$16 - 9 = \_\_\_$

Thank You God for blessing us with math and how it gives us different ways to explore and understand Your creation!

Hi, I'm Dr. Diana Carle, the writer and illustrator behind this book, and it is a complete joy to share the wonder of learning in God's creation with you! I earned my PhD in Entomology, a Bachelor of Science in Biology and Anthropology, as well as a minor in the Classics. However, my most impactful educational experience was the nature-based homeschool program in which I grew up. As I explored and sketched the intricacies of life along the pond's edge or studied the miracle of butterfly metamorphosis, I learned at a young age that no classroom was as captivating or rooted in truth as the one created by God in nature. Professionally, I have spent over 13 years in scientific research and STEM education, creating and delivering a wide range of hands-on science-based curricula for undergraduate up to graduate and professional education courses. I also tutor K-12 students in English, math, art, and the sciences while creating storybooks and educational materials. Most importantly though, I continue to learn from God in His creation through stewarding the life in my own backyard.

www.DoctorDianaCarle.com

Thank you so much for reading, I hope you experienced some of the wonder that learning in nature holds. If you are inspired to explore more about the natural world around us and how to use it as a learning resource, check out my other books, and join the mailing list by emailing me at:

DrDianaCarle@gmail.com

To Get Free Extras!

Title the email,
"Nature's Living Classroom"
And I'll send you a special guide with activities for learning in nature.

Made in the USA
Monee, IL
07 November 2026

67873761R20077